AF681436

Special Operations and Elite Units, 1939–1988

Recent Titles in
Research Guides in Military Studies
Series Adviser: Roger J. Spiller

The Peacetime Army, 1900-1941: A Research Guide
Marvin Fletcher

Special Operations and Elite Units, 1939-1988: A Research Guide
Roger Beaumont

Special Operations and Elite Units, 1939–1988

A Research Guide

Roger Beaumont

Research Guides in Military Studies, Number 2

Greenwood Press
New York • Westport, Connecticut • London

Library of Congress Cataloging-in-Publication Data

Beaumont, Roger A.
Special operations and elite units, 1939-1988 : a research guide / Roger Beaumont.
p. cm. — (Research guides in military studies, ISSN 0899-0166 ; no. 2)
Bibliography: p.
Includes indexes.
ISBN 0-313-26001-X (lib. bdg. : alk. paper)
1. Special forces (Military science)—Bibliography. I. Title. II. Series.
Z6724.S66B4 1988
[U262]
016.3553'5—dc19 88-25083

British Library Cataloguing in Publication Data is available.

Copyright © 1988 by Roger Beaumont

All rights reserved. No portion of this book may be reproduced, by any process or technique, without the express written consent of the publisher.

Library of Congress Catalog Card Number: 88-25083
ISBN: 0-313-26001-X
ISSN: 0899-0166

First published in 1988

Greenwood Press, Inc.
88 Post Road West, Westport, Connecticut 06881

Printed in the United States of America

The paper used in this book complies with the Permanent Paper Standard issued by the National Information Standards Organization (Z39.48-1984).

10 9 8 7 6 5 4 3 2 1

To my Aunt and Uncle

The late Marguerite Poser Neelen

and

The Honorable Harvey L. Neelen

Always in my corner

Contents

SERIES FOREWORD

In modern military parlance, "Special Operations" denotes an unconventional military action, often covert, which is usually carried out by selectively recruited, intensely trained, and specially equipped forces for strictly defined objectives of strategic, operational, or tactical consequence. And while war in the 20th Century has slipped the formal confinements of its traditional definition, allowing for an expanding role of regular forces in times of peace, Special Operations have not been governed by the normal conventions of conflict between nation-states. The Joint Chiefs of Staff have made explicit the distinction between regular and Special Operations in their *Dictionary of Military and Associated Terms*:

> [Special] Operations [are] conducted by specially trained, equipped, and organized DOD [Department of Defense] forces against strategic or tactical targets in pursuit of national military, political, or psychological objectives. These operations may be conducted during periods of peace or hostilities. They may support conventional operations, or they may be prosecuted independently when the use of conventional forces is either inappropriate or infeasible.

Roger Beaumont's *Special Operations and Elite Units*, the newest entry in Greenwood Press' *Research Guides in Military Studies*, addresses a unique, little-known, and commonly misunderstood facet of modern military practice, one for which national leaders and their military organizations have of late shown an improved (if, as in the United States, a somewhat equivocal) appetite. The nature of conflict in this century has veered away from the large-scale conventional operations that characterized the two World Wars toward hostilities with a much lower order of magnitude whose objectives are more highly circumscribed. The tempos and rhythms of conflict have been altered corre-spondingly. Guerrilla war, revolutionary and counter-revolutionary war, as well as trans-national terrorism, protract and telescope violence without heed to time or even, in some cases, location. The ambiguous and episodic nature of modern hostilities has created an environment to which Special Operations seem particularly well suited.

Yet, as Professor Beaumont explains in his introduction, Special Operations have often held out promises they could not keep, the forces conducting Special Operations have often been misused, and in general have often inspired a good deal of wishful thinking and self-delusion by their masters and the forces themselves. Innately attractive because of their exclusive character, because they are ordinarily kept at a high state of readiness, and because of their relatively limited size, Special Operating forces seem handy and precisely controllable to those who exercise authority over them. Too, the secretiveness with which such forces usually operate has always been appropriate to highly sensitive political and military missions, missions that carry with them consequences far outweighing the forces actually involved if the mission misfires.

Professor Beaumont has produced what may well be the single most comprehensive survey in any language of modern Special Operations literature. His brilliant and wide-ranging introduction sets the subject in its historical, typological, and national contexts, and his bibliographic investigations are arranged for the benefit of readers new to the subject. Scholars and students of military affairs, government officials--as well as practitioners of Special Operations themselves--should find a great deal of interest in this volume.

Roger J. Spiller
U.S. Army Command and General
Staff College
Ft. Leavenworth, Kansas

Preface

Like any bibliography on a complex subject beyond an alphabetical listing, this research guide is subjective and organized arbitrarily. A broad sampling of material is presented, most of it accessible through interlibrary loan services to those far from central archives and major research libraries. Much material has been excluded and the purpose of inclusion has often been illustrative. A full compilation might well run to several volumes. Some may find that important in itself. Others may question the inclusion of psychological operations with special operations. That is also done deliberately, in keeping with the original fusion brought back into focus recently in American defense circles by Al Paddock.

It should be noted at the outset that while many *corps d'élite* were created long before 1939, the real flowering of such units and special operations came in the Second World War. At the same time, special operations have been carried out on occasion by other than *corps d'élite*.

In the essay, I have covered some of the same ground as I did in *Military Elites* over a decade ago. Much has changed since then although the essence, problems, and paradoxes remain the same. Ignoring that fact has cost much along the way. The appendices are designed to offer perspective and to reflect a mushrooming phenomenon.

There is not enough space here to deal with such unfolding problems as police SWAT teams; psychological dimensions of creating, selecting, and employing *corps d'élite*; or dependency on or stylistic preference for, or revulsion toward, special units and operations.

In respect to categorization, there is often a tendency, stemming from their small size and light weaponry, to see special units and operations as "tactical." Nevertheless, they have often been strategic in the consequence of their acts, in their political flavor, and in their symbolism as a kind of deterrence--often far more than the units or their controllers sense. Yet there is little concern in "open" analysis about such aspects or about their potential as a catalyst to escalation, either through action, serving as a symbol inadvertently, or through misreading the degree of impact, such as being ambushed or captured in a crisis setting.

There are many areas, therefore, in which more research needs to be done in this complex realm in which, for some time to come, the boundary between "academic" and "real-world" may not be as sharply drawn as it seems at first glance.

Acknowledgements:

Thanks are due to the following for their aid in pursuing this project:

Lieutenant Colonel Macdonald Valentine, USAR, Atlanta, Georgia

Dr. John Greenwood, Chief Historian, Corps of Engineers, Baltimore, Maryland

Colonel Juan S. Eleta, Office of the Military Attaché, Embassy of Spain, Washington, D.C.

Dr. Dean Allard, Historical Division, United States Navy, Washington, D.C.

Warrant Officer 1 K. A. Fowler, British Defence Staff, Washington, D.C.

Colonel Hussenot-Desenonges, Assistant Military Attaché, Embassy of France, Washington, D.C.

Master Sergeant Claus Munsch, Office of the Military Attaché, Embassy of the Federal Republic of Germany, Washington, D.C.

Lieutenant Colonel P. H. Pedersen, Royal Danish Air Force, of the Defense, Military, Naval and Air Attaché's Office, Royal Danish Embassy, Washington, D.C.

Master Warrant Officer R. P. Sweeney, Canadian Defense Forces, of the Canadian Defense Liaison Staff, Washington, D.C.

Commander B. A. Van Dijk, Royal Netherlands Navy, Assistant Naval Attaché, Royal Netherlands Embassy, Washington, D.C.

The *Bundesminister* of the Interior, Bonn, especially Herr Renz

The staff of Texas A&M University Library's Interlibrary Loan, especially Angela Crik, Sean Cleveland, Jackie Halverson, Susan Scott, and Matt Krzysiak

Texas A&M History Department staff: Deborah Callihan, Doinan Kavanaugh, Merle Rucker, and Jude Swank;

My wife Penny and daughter Anne for aiding in copy-editing.

Special thanks are due Laura De Sola, Gary Sinton and Dean Carl Erdman of the Texas A&M Engineering Program whose special aid and cooperation made completion of the project possible;

And I am grateful to Larry Hill, Head of the Texas A&M History Department, and Associate Dean Arnold Vedlitz of the College of Liberal Arts for supporting research in the summer of 1987 which allowed me to get a fine grasp on the project.

Prologue

Imagine you are a lieutenant general in the United States Army in early 1942. A career officer for twenty-two years, you have vaulted forward five ranks in six months. After a generation of lean budgets and frequent insults, you find that the nation's eyes are on you and your comrades in the professional officer corps to save the day. Everywhere on the map above your desk are tell-tale red arrows and dotted lines of enemy victories, symbolizing the disintegration of your own forces, the fruits of years of peacetime parsimony. You are making decisions for ten, twelve, sixteen, sometimes twenty hours a day, under often contradictory pressures from the field, from above and from politicians, once champions of isolation and thrift, who now champion the war effort. Gnawing doubts mount as fatigue increases, and the pile of paper in your in-basket spills over. You are wearing glasses for the first time in your life. You have insomnia, and your waist thickens as your day is eaten up by meetings. Your domestic life is becoming a casualty of your seven-day week, since the national emergency is not taken quite so seriously at home.

Underneath the vast and confused flow of communication is a vital pattern. Your decisions about men and resources will mark the pages of history. The fate of your country and much of the world is at stake. You cannot foresee all the long-term effects of what you decide to do or not to do in a hurry or upon impulse. As a professional, you know you are designing not only a military system for war but one for peace that follows. For when the war ends, win or lose, you may once again be ignored and slighted, then later criticized by historians, comfortable in hindsight and at leisure in quiet offices and carrels, or, at short range, beset by journalists or politicians.

And you know that, while many with you in the struggle strive to be selfless and dedicated, there are also some who see war as a way to get government contracts and milk them, who may make their way undetected, or even leave you tainted with scandal. Within the service ambition also flourishes, aimed not at money, but power and reputation. Some will move up the promotion scale by doing something dashing, not lacking bravery, but sometimes all too eager to clamber up the ladder of glory and promotion.

How can you protect your country and yourself against all this? Here, for example, is a proposal, atop this morning's pile of "immediate action" documents, bucked down from the Chief of Staff for your decision. Young Colonel X of the XXXth Infantry wants to form a special battalion of cliffscalers, made up of mountaineer and spelunker volunteers from the whole army. He argues that because regular divisions tend to be road-bound, his unique unit is needed to move easily across broken terrain, restoring initiative to our retreating forces, beating the enemy in an area where his great initial strength and initiative cannot be brought to bear.

The proposal includes a tissue paper overlay and world map marked to show lava extrusions and rough roadless terrain where units of this rough-terrain corps could move against enemy flanks, harassing and striking terror. Give me three months, the young colonel says, and I'll win at least a token victory to hearten the whole nation. As we draw the motorized enemy into this rough, forbidding terrain, the individual fighting man will again be restored to prominence, and machines offset. Later, the Rough Terrain Training Center will generate a series of such units, and be a seed-bed of leadership for the army, instilling the infantry and other branches with a zeal to surprise and to conquer tough obstacles, instead of plodding along the road into ambush, or slogging away in attrition combat.

The tightly-written prose flows well. The young colonel is articulate and his enthusiasm moderated. He has done his homework. He knows his field. He is bold and ready to come to grips with the enemy.

You yawn and look at your watch. Six-thirty p.m. You haven't had dinner yet. But what about this Rough Terrain Corps proposal? You switch on your dictating machine.

"From Assistant Chief of Staff, Plans and Development, to Chief of Staff. Subject: Memo, Colonel X to Chief of Staff . . . Miss Jones, take the date off the memo, please . . . This is about the twelfth high-powered peanut stand proposal this month. I appreciate there's nothing duller than training and retraining and itching for some action, and I know his boss, the CO of the XXXth. He's hard to live with at close quarters. Nevertheless, gathering all the talent from the general forces that Colonel X wants under his wing would require a special table of organization. Even if they captured Hitler or Mussolini or Tojo, I'm not sure it would be worth it in terms of the effect on general morale. Hell, every combat soldier should be qualified for this RTC. If we create one, some folks will see it as an area where they can slack off in training. It could even make us more roadbound overall in the end. I would reject this proposal on the grounds that training space and resources for such a program are not available under the current dire emergency. . . ."

You switch off the dictaphone, and head for dinner. As you enter the elevator, you think of all the young colonels all over the world who want to do something different, something exciting. You think about what a colorless, exhausting business preparing an army is for most of the people doing it, and for you. At least Colonel X is out in the field with troops. What if he were locked up in this super-dungeon of the Pentagon wading through piles of paper, with no chance ever for field service? Then, you think, hell, maybe I made the wrong decision. At least if they rounded up lots of these fireballs and oddballs and got them away someplace working hard at climbing hills, it would make life easier for the MPs. A tough call. A very tough call. . . .

Special Operations and Elite Units, 1939–1988

Introduction

Why have elite forces units flourished against the tides of mass culture and bureaucratization of the military? How much have they reflected or contradicted their parent system's values? How often have *corps d'élite* matched their creators' hopes? Has their performance justified immunity from orthodox control and the granting of special resources and adulation?

Before examining those questions, others must be asked. First, what is an elite unit? There is no simple one-sentence definition. Many of those created in the twentieth century were deemed elite before actual formation. Physical and mental standards for admission have been high, and a special premium placed on stamina and quick reaction. They have had distinctive uniforms and equipment or insignia and unique traditions and customs--often synthetic and bizarre. Elite forces have also often been relatively free from ordinary military control and discipline. Acceptance has usually involved surviving an ordeal, a "rite of passage," requiring tolerance for pain, discomfort, and risk. Many *corps d'élite*--but not all--have received much publicity. In many cases, they have acted out disdain for outsiders, their own headquarters and support personnel, civilians, and even for adjacent allied combat units, sometimes more so than the enemy.

And what are "special operations?" Definitions are often vague, but the term has been used to include raids, reconaissance, capture or theft of equipment and kidnapping in peace and war, formation, training and liaison with local irregular or insurgent forces and counter-terror operations, demolitions, sabotage and assasination. It should be noted that special operations have been carried out by standard forces, as in the shooting down of Admiral Isoruku Yamamoto in April, 1943, and in the Hammelburg raid, while special units have often been assigned roles in war or low-intensity conflict far from what they had been trained for or indistinguishable from the missions of conventional forces.

Modern military elite units have frequently been born of frustration, created in response to enemy success, and loss of initiative. They usually trade youthful daring for time and resources, sometimes trying to make up for deficiencies revealed in combat. Many of the special units created since World War I have been copies of successful *corps d'élite*. At the same time, in spite of their heroic image, membership in an elite force has often not been "career enhancing."

Paradoxically, creating *corps d'élite* has often failed to solve the problem perceived at their founding, and sometimes they have created new ones. Not only has their special status produced intra-organization friction, but also their popular images have often obscured unpleasant side effects. Military history is often written in combative cliches, focusing selectively on movement of units and the clash of personalities, and overlooking more complex dynamics and problems.[1]

Since the decision to create such forces was often made in haste and in reaction to a specific challenge, elite forces have often found conditions radically changed by the time they were ready for action. They have also frequently been used--and misused--for other purposes. Ground force *corps d'élite*, being lightly equipped and usually formed to be hit-and-run raiders, suffered heavily when committed to sustained combat.

Their basic purposes have varied enough to suggest a typology:

Types	Examples
Ceremonial/social regiment	U.S. Army 3rd Infantry Regiment; Honorable Artillery Company
Traditional--combat-proven over time	Guards regiments of various nations; French Foreign Legion"Post-hoc"--combat--proven in a specific conflict
"Post-hoc"--combat--proven in a specific conflict	The Iron Brigade in the Civil War; the "Fighting 69th" in World War I; the U.S. 1st Infantry Division in World War II
Praetorian (bodyguard)	The *SS Leibstandarte*; Papal Guard
Ethnic/cultural	Gurkhas; Cossacks
Politico--ideological	International Brigades of the Spanish Civil War; NKVD Units of the Red Army
Romanticist-predesignated	Commandos/Rangers/U.S. Marine Raiders
Technological/cybernetic	79th Armoured Division; RAF Pathfinders; "Old Crows"
Nihilistic/mystical	Spanish Foreign Legion; *kamikazes*
Specifically functional	Airborne; aviators; submarines

Obviously, some types would fall into more than one category, e.g., the 2nd Marine Raider Battalion was functional, nihilistic, romanticist, and ideological. *Corps d'élite* in existence at the beginning of the twentieth century were of the first four types. Those formed after 1920 are generally found in the bottom half of the list, although many *corps d'élite* in the Third World have had a Praetorian function.

A reasonable person pondering things over on New Year's Eve, 1899, might have expected *corps d'élite* formed in the new century to be less romantic than their predecessors, e.g., Guards regiments and "fashionable" units, but to be rather more like ships' engineers who, however vital, did not share the lofty status of Navy line officers. Throughout the nineteenth century, functional auxiliary units had been refined or created, such as mountain troops, signal and railway corps, fog troops, and expanded medical services. However, most twentieth-century *corps d'élite* did not merely supplement fighting services but actually upstaged them. Even when such forces were created with technological change in mind, their glamour derived as much from *mystique* rather than from technique.

The twentieth century, of course, has seen a widespread shattering of social stability. Aristocracy had been crippled in the eighteenth century in France by guillotining and in America by the musket, and in the nineteenth, in Britain by the ballot, but in the twentieth century, traditional privileged castes were swept from power throughout the world. Nation after nation came under the rule of new leaders, far from the athletic, eccentric nobility as bourgeois values triumphed in the democratic West and the autocratic East. The business-suited bureaucrat became a symbol of communism, just as business-suited politicians, managers, and bureaucrats--military and civilian--gained authority in the West. In peacetime, soldiers and sailors wore mufti in Washington and London to minimize visibility in peacetime years of budget trimming.

While civilians more and more usurped military authority in Western liberal democracies, in many other parts of the world, post-colonial disorganization and political instability in the wake of the dismantling of aristocracies generated a pandemic of dictatorships. In democracies and dictatorships, as upper echelons grew more bureaucratic, and noncombat support elements proliferated, *corps d'élite* became an escape for the young and adventurous, analogous to a form of sport. Since military doctrine and structures often failed to reflect how young men more and more actually fought wars, well away from central authority, elite forces often seemed a way to escape from humdrum, desk-bound, careerist, harem-like "small-p" politics, from committee-ridden anonymity and consensus, and from old-boy dominance. And not infrequently, at the upper peak of the organization pyramid, some saw special units as a way to keep Young Turks busy.

To critics of mass culture, elite units might seem a manifestation of a hunger for heroes in an "Age of Anxiety," where millions lead "lives of quiet desperation," torn from the soil, victims of blind forces and heedless cruelties, debased physically and mentally by repetitive and meaningless work and synthetic food, mere spectators of tawdry thrills. Elite forces, in such a context of *anomie*, symbolize individual courage triumphant in the face of mass destruction, mechanization and mediocrity, themes manifest in the cinema and the comic strip, where crude synthetic popular heroes have abounded since the 1930s. It is not hard to argue that the romantic heroism of nineteenth century literature and arts was debased and extended in the age of mass communication. Hitler, a devotee of Wagner, harnessed that trend as he laid forth a scheme for using the image of combat *corps d'élite* in *Mein Kampf*.

Disconcertingly for our rationally-minded prophet of 1900, the impact of standardized industrial technology and other collectivizing pressures was uneven, although grim in its toll when the First World War brought those forces into a massive collision, then stalemate, and ponderous grinding battles as massed artillery, machine guns, and barbed wire dotted vast moonscapes of devastation on several fronts.

Several armies responded to the stagnation by resorting to elitism, from which a new and perverse romanticism bloomed, deemed "heroic realism" by Ernst Jünger in its German variant. In the air, and in the battles of 1918, a new pattern of war resembling a medieval tournament appeared, one which endured into the Second World War. Many watched and supported, while a few--often, as in the Battle of Britain, the very few--fought. In the Western democracies' armies, patterned after the division of labor found in industrial systems, many became clerks, teamsters, freight handlers, cooks, and so forth, the average line infantry units consisting mainly of men without personal skills or social leverage to avoid them.

In World War II, some five million of sixteen million Americans who entered the service went overseas. In that war, battles of the scale of the First World War occurred mainly on the Russian front. In the West, more and more, infantry became staked goats with artillery and fighter-bombers as the hunters. S.L.A. Marshall created a minor sensation in 1948 when, in *Men Against Fire*, he purported that U.S. Army military historians' battlefield interviews in World War II showed that "less than twenty-five percent of men in combat . . . fired their weapons at all" and that only fifteen percent participated in fire-fights on a regular basis.

Although some attempts at reform were undertaken, the infantry continued to be seen as the main battle element in modern battle, although artillery had done most of the butcher's work in both world wars, and in Korea as well. Throughout the nineteenth century and into the First World War, infantry had been forced to ever greater dispersion as the Confederates' Enfields, the Boers' Mausers, and the Kaiser's machine guns cut swaths through swarms of men. Mass and observer-adjusted indirect artillery fire and machine guns drove foot-soldiers to ground and, from the 1930s onward, air power amplified the effect. Yet acceptance of such realities into military organization came fitfully, and many civilians still view combat as a kind of mob scene, partly due to graphic compression in film and video.

Viewing *corps d'élite* mechanistically, however, as rational tactical instruments designed in reaction to dispersion or systemic analysis misses the fact that the creation of *corps d'élite* has often been reactive under stress and may therefore reveal more about their parent societies than their creators intended. One aspect of their proliferation taken for granted is the development of "objective" physical and mental standards by doctors and social scientists which allowed a "screening" of the general "pool" of military manpower. The sense of elitism thus reinforced by science paralleled the fusion of eugenics and genetics (and pseudo versions thereof) with politics, often reinforcing a sense of elitism in a special and perverse way. Those who survived screening and *corps d'élite* entry ordeals and battle have often gained a special sense of importance and achievement, feeling tough, heedless of death or even rash, favored by fate, and superior to most forces on their own side. Refusal of many American airborne enlisted men to salute "leg" (non-airborne-qualified) officers reflected that sense of specialty. In the Second World War and after, many commanders, military police, and other units came to resent the truculence and swagger of elite forces.

Changes in weapons technology also led to the growing reliance on elite forces in low-intensity conflict. In the early twentieth century, rifles, machine guns, automatic pistols, submachine guns, new explosives and radios--and the mimeograph machine--entered the guerrilla and terrorist arsenal as mass education and media extended and intensified the fever of nationalism. Insurgents could take greater advantage of the sluggish reflexes of mass forces, strike harder with less risk, and reach out to the masses. In the First World War, in Arabia-Palestine, Ireland, and East Africa, tiny bands tied up vast armies. In the 1920s and 1930s, the Irish Troubles, the Rif revolt in Morocco and Algeria, the Russian Civil War, the Syrian revolt, Spain, China, and the Gran Chaco War all marked the proliferation of guerrilla war, while the rise of the Congress Party, Fascism, Nazism, and the new Soviet Man showed a rise in sophisticated manipulation of mass psychology.

The Second World War mortally wounded Western imperialism by revealing the colonial powers' military flaws and by providing sophisticating experiences and guns to subject peoples. After World War II, elite forces became prime instruments in fighting insurgency. Although often functioning as virtual police, unhappily and in vain *corps d'élite* tended to seek personal combat as though involved in conventional war, and fighting--or trying to fight--insurgents frequently led to frustration and atrocity. The hunger of some elite forces for a stand-up fight became a point of leverage for terrorists and guerrillas. If they could be goaded into excess, it would yield sympathy and support for insurgent causes. The paradox that much of the world expected a double standard of behavior in "limited wars" made little difference.

Premier Nikita Khrushchev's pronouncement of Soviet support for "wars of national liberation" in 1956 came in the wake of the frustration of Chinese and North Korean designs in Korea by American firepower--one million Chinese and North Korean dead versus 33,000 Americans, stalemate and partition in Indochina, and defeat of Communist insurgents in the Philippines, Malaya, and Greece. For the most part, those costly confrontations to varying degrees underlined the bitter lesson learned by Japan in fully arousing American ire. Although new model of insurgency would be more ephemeral, slower and while not always successful, it would, for the most part, avoid prolonged exposure to Western firepower and linear stand-up battles, except when advantage seemed clear as on the Suez Canal in 1973 and in South Vietnam in 1975. The response, then, to the gossamer tactics of insurgency was the lightest of light forces--the new variants of the *corps d'élite* which appeared in profusion in World War II.

If World War I's stormtroops had been the prototype of modern *corps d'élite*, World War II was an assembly line as literally dozens of special units were formed. (See Appendix B.) In 1939, the German intelligence service deployed teams known as the Brandenburgers, who often wore the other side's uniforms as they disrupted, deceived, and gathered intelligence. Allied *corps d'élite* were designed for raids like the French Army's *Corps-franc* and the British Independent Companies, forerunners of the Commandos formed in 1940.

British and later American "mobs for jobs" symbolized individual aggressiveness, compensated for early defeats, and made up for weakness in other areas. Once the Allies regained the initiative, many were converted into conventional units or disbanded. No new models appeared after 1943, and only a few of the species survived the war as existing organizations--the Commandos, British and Belgian Special Air Service, and various Allied airborne and underwater units. "Mobs for jobs" had first made their weight felt seriously in 1941-42 in North Africa, when British "mobs for jobs" tore up the rear of Erwin Rommel's Afrika Korps, wiping out hundreds of planes and vehicles and raising general hell. The Axis powers were forced to divert dozens of divisions to fight Allied-supported partisans in the Balkans and behind the lines in Russia and to fend off Commando raids and Resistance activities in Western Europe. In the Far East, a whole garden of Allied special units sprang up as well. In 1943, Hitler, impressed, ordered an expansion of the Brandenburgers and an increase in operations.

Thus it was that a few thousand clean-cut, enthusiastic youths on both sides were trained to slit throats, gouge eyes, break backs, and crush testicles in hand-to-hand fighting, showing remarkable enthusiasm for doing things that if done in another setting would have led to a death or felony sentence and/or the revulsion of the community back home. Although many military professionals saw such activities as outside the ethics of soldiering, the public on both sides lapped up the exploits of special units, even when lurid details were made known.

Whatever advantage they gained, like submarine attacks on merchantmen and the bombing of cities, special units undermined the ideal that soldiers were agents of the state under close obligation to play by rules, and a special code of honor. During World War II, and after, as the "mobs for jobs" revitalized surprise as a principle of war, they sometimes surprised and embarrassed their own side.

At the same time, creating elite units offered a way for higher commanders to more closely control tactical units and, in conscript armies, to form high-quality units per se, even though, paradoxically, their existence suggested that standard forces were not quite up to deft, fast-moving operations. They were also seen as a nuisance by many commanders ordered to provide them support and administration. Some, like the Australian Z Special Force and the British Force 136, 204 Mission, V-Force, and Operations Centre Khartoum, were closer to intelligence-sabotage networks than military combat operations. The most visible ground-force elites were light infantry in organization and weapons. In the First World War, storm troop and tank enthusiasts had visualized armies as organisms, with delicate nerves and circulatory systems behind a relatively thin skin, and elite forces as scalpels to wield against enemy lines of communication and critical points. But there was a difference; storm troops and tanks in the "Great War" penetrated the enemy front in coordination with major conventional following elements, while "mobs for jobs" often operated as harassers and freebooters, unorchestrated and anomalous.

Beyond that lay another paradox. Because the high quality and zeal of the elite forces' personnel stood in contrast to a shortage of manpower in parent forces, many *corps d'élite* were involved in prolonged fighting against standard units with heavier weapons and better support facilities. Aside from the resultant heavy losses, there was the question of a "leadership drain." How much did special units rob their parent organizations of leadership? Should not the privates in the U.S. Marine Corps, the air forces, or the Royal Marine Commandos in World War II have been sergeants and lieutenants in the conscript army?

Scanning the titles and appendices offers a sense of the range and complexity of such academic and policy questions, and the scale of the proliferation of *corps d'élite* and special operations, as well as aiding the researcher bent on a particular focus. The following general overview of evolution since 1939 will also provide perspective, and it is appropriate to begin with perhaps the most widely copied form of special unit, the British Commandos of World War II.

THE COMMANDOS

The World War II Commandos were born of Winston Churchill's combative enthusiasm in the dark days of 1940. As with the airborne, Churchill confronted War Office inertia and resistance from commanders, after calling for a rapid-moving, aggressive corps of "storm troops" even before the 1940 campaign in France had ended. Contemplating arrival of Australian troops later diverted to the Middle East, he suggested a fire-brigade role for them, with light vehicles and automatic weapons, to rapidly counter invasion and eventually to raid the Continent.

When a staff officer, Lieutenant Colonel Dudley Clarke, proposed a volunteer raiding force, Churchill demanded 10,000 such "leopards," citing the example of German storm troops in 1918. The Commandos were the result, their name adapted from Boer mounted guerrilla bands in the South African War of Churchill's youth. As invasion loomed in the summer of 1940, Churchill and Commando leaders battled opponents who feared loss of the army's best men and weakened morale if special attack units were formed. The Commandos survived, their goals being to kindle the spirit of the attack, gain battle experience, keep the Germans off balance, to put Britain on the offensive, and to serve as a laboratory of "triphibious tactics."

Formed around twelve "Independent Companies" of Territorial Army volunteers raised in early 1940 for special operations, the fledgling Commandos, after intensive training in field craft, small-unit tactics, and raiding, raided Boulogne in late June and Guernsey in mid-July 1940, using high-speed motorboats. The first commander of Combined Operations--the headquarters controlling the Commandos and their air and naval support--was Admiral Sir Roger Keyes, V.C. Also facing bureaucratic obstruction, Keyes held out for large-scale operations to make a large publicity splash. But, as described in the novels of Commando veterans Robert Henriques and Evelyn Waugh, inactivity weakened morale. After the Löfoten and Spitzbergen operations in the fall of 1941, the latter conducted by Canadian troops, Admiral (later Admiral of the Fleet) Lord Louis Mountbatten replaced Keyes. His youthfulness and drive were part of a British effort to substitute grander raids for the full-scale "Second Front" invasion in 1942 urged by the Americans to aid the Russians but to be conducted by British forces. Mountbatten's commando "empire" expanded rapidly. Another large raid, again in the Löfoten Islands off Norway, came in late December and was followed by others against the French coast and on Crete, and in Syria and North Africa in early 1942. Although the large, bloody "raid" at Dieppe in August 1942 was under Combined Operations, few Commandos were involved. Eventually, 25,000 Commandos saw action in Madagascar (1942), Tunisia (1942-43), Burma (1945), and Germany (1945), but more and more fielded as infantry in sustained combat, rather than as raiding forces. Immediately after the war, much reduced Commando forces were maintained in Britain, the Commonwealth, and in Belgium.

On balance, they were a success. The Commandos symbolized Britain fighting against Nazism. Military discipline may have been aided in the pre-invasion period by attracting aggressive volunteers into remote areas for mountain and amphibious training. Beyond that, they caused the Germans substantial concern and raised Hitler to both fury and emulation. Since the British Commandos generated no political virulence later, it is easy to give Colonel Clarke and Churchill high marks. But there was a debit side to the ledger as well. Even short, sharp contacts with the enemy exacted a price above Western Allied casualty norms in World War II; 1,760 Commandos of 25,000 trained were killed in action, although all did not see combat, or did so for very brief periods of time. But perhaps most significantly, the Commando concept set a pattern for a new kind of war which fell back from standards evolving since the seventeenth century. Although the Nazis were hardly credible as a source of moralistic condemnation and they adopted the model, their depiction of Commandos as a kind of gangster in uniform was not wholly off the mark. Since World War II, the term has become a universal rubric for elite raiding forces, and a great many imitations created with varying degrees of conformity to the rigor of the prototype in selection and training.

THE RANGERS

The Americans also fell in step. From December 1941 to November 1942, as the United States, like the British earlier, suffered humiliating defeats and were unable to come to grips with the enemy on a large scale, the Commando model of aggressiveness was naturally attractive. In June 1942, Ranger companies based on the Commando model were raised from U.S. troops stationed in Northern Ireland, a concept approved by General George C. Marshall, U.S. Army Chief of Staff, at the suggestion of Brigadier General Lucian Truscott, head of a U.S. Mission to Combined Operations. (While the "Rangers" title was probably derived from the film version of Kenneth Roberts' novel *Northwest Passage*--few then or since noted that Rogers' Rangers served the Crown in the Revolution.)

After training with the Commandos at Achnacurry, baptism of fire came in August 1942 at Dieppe. Although Canadian infantry units were the bulk of the force and suffered most of the casualties and the Rangers returned unscathed, American newspapers suggested a stronger American contingent at Dieppe than was the case. It was a lucky beginning but, in time, the Rangers paid a steep bill for elite status.

A full battalion of Rangers commanded by Lieutenant Colonel William O. Darby was created and served in North Africa in 1942. By early 1943, "Darby's Rangers" had grown to a light brigade and two battalions went ashore at Gela on Sicily in July. A special task force under Darby's command in Patton's drive on Palermo captured almost 4,000 prisoners in one day. Still their luck held.

In the late summer, the Rangers landed near Salerno and were involved in the drive on Naples. Although selected, trained, and lightly equipped for hit-and-run roles, the Rangers more and more played a conventional infantry role. In January 1944, three Ranger battalions took Anzio with relative ease. Ordered to advance to the east at night to spearhead the 3rd U.S. Infantry Division's attack on Cisterna, the 1st and 3rd Ranger Battalion, creeping slowly along a flooded irrigation ditch at dawn on January 30, 1944, were ambushed by German tanks and self-propelled guns at close quarters. By noon, half of the Rangers were dead or seriously wounded. English-speaking German officers carried hand loudspeakers, using hostages to force others to surrender. Although ten percent of the captured Rangers later escaped, two battalions were wiped out. Darby tried to reach the beleaguered force with the 4th Battalion. After suffering fifty percent casualties and death of all its company commanders, the 4th fell back. Of 767 Rangers who moved against Cisterna on January 29, six returned.

Of an original 1,500, only 449 Rangers remained. Two hundred and fifty were transferred to the 1st Special Service Force (see below) and the rest sent home to train troops. Darby, promoted to brigadier general and assistant division commander of the 10th Mountain Division, was killed in action in early 1945. The remaining Ranger battalions--six were formed in all--fought on. In the Philippines, the 6th participated in a spectacular raid, freeing American prisoners of war at Cabanatuan. On D-Day, the 2nd and 5th Battalions lost 197 out of 355 men who landed.[2] (The Kachin Rangers, a term applying to OSS-led local forces in Burma were not Rangers within the U.S. Army structure.)

After the Second World War, the Ranger concept lived on unevenly in the U.S. Army until the late 1960s. Some American divisions formed "Ranger" units, and battalions were organized in Korea. In the mid-1950s, however, Ranger units were disbanded. Special training of individual volunteers at the Ranger School led to the award of Ranger insignia. The term was used in 1968 when the U.S. Army formed a provisional battalion, which grew in stages to the current Ranger Regiment of three battalions, elements of which were involved in Operation *Urgent Fury*, the U.S. invasion of Grenada in 1983.

THE FIRST SPECIAL SERVICE FORCE (FSSF)

The U.S.-Canadian 1st Special Service Force, created for a mission which quickly evaporated, was a bizarre byproduct of Anglo-American strategy formulated at the Quebec Conference. With the support of Admiral Mountbatten, chief of Combined Operations, a scientist, Geoffrey Pyke (who also suggested huge aircraft carriers made of "Pykrete," a mixture of sawdust and ice), proposed using highly mobile tracked vehicles to destroy Norwegian industry.

To carry out the mission, United States and Canadian complements were merged into the FSSF[3] at Fort William Henry Harrison near Helena, Montana, under the command of a young American officer seen as a "comer," Lieutenant Colonel Robert Frederick. While the Canadian half was hand-picked, the American half was culled from stockades and included unit rejects, many testing low on the U.S. Army intelligence scales. After extensive reassignments, Frederick, travelling regularly to Washington to fight bureaucratic battles, shaped the 1st into his own creation. The unit wore red berets, and U.S. Army Indian Scouts' crossed arrows as its insignia--also adopted by U.S. Army Special Forces when they became a branch in 1986.

Although it underwent airborne, mountain, and arctic training, the unit never jumped in combat or operated in high mountain or snow country. When the FSSF faced disbandment after the Norwegian government-in-exile vetoed Pyke's scheme for destroying Norway's economy, Frederick fended off dissolution. The 1st was committed--without its "Weasel" vehicles--to the Aleutians campaign, then on the Winter Line in Italy (at Anzio and Rome), and finally in the Rhone Valley. Having suffered high casualties and without a special role, the 1st was disbanded in December 1944, most survivors assigned to airborne units.

Later claimed as a triumph of U.S.-Canadian relations, the FSSF was the focus of an "inordinate amount of high level consideration."[4] Placing Canadians under U.S. command generated extensive legislation in Canada's parliament. The senior Canadian officer communicated directly with Ottawa and the Combined Chiefs of Staff, and U.S. and Canadian diplomats also spent much time discussing the unit. Outside British Army channels in Europe, the 1st had to petition for British decorations.[5] The symbolic value of the "Devil's Brigade" and Frederick's charisma had kept the unit alive well beyond its utility.

THE LONG RANGE DESERT GROUP (LRDG)

In the Second World War, *corps d'élite* abounded in North Africa and the Far East as well. Contenders for Clive of India's and Lawrence of Arabia's laurels were not lacking. The first major Allied special unit formed outside Europe was the Long Range Desert Group (LRDG). In late 1940, British Empire forces in North Africa numbered less than 20,000 combat effectives against 300,000 Italians. Before the war, dry rot in Fascist militarism had been visible in Ethiopia, Spain, and Albania, and in June 1940, six French mountain divisions repulsed twenty-four Italian divisions. In Egypt and Libya in 1940, as odds seemed overwhelmingly one-sided and British air power was stretched thin, lack of effective air reconnaissance led General Archibald P. Wavell, then British Middle East commander, to approve Brigadier Ralph Bagnold's request to form a special ground reconnaissance force. Bagnold, a Royal

Engineer officer and brother of Enid Bagnold, author of *National Velvet*, had explored North African deserts in the 1920s and 1930s and was a noted academic expert on sand movement and terrain. Thirty New Zealanders formed the core of the Long Range Desert Group, which grew to include three hundred volunteers from Rhodesian and British units and twelve Americans.

In over two hundred operations, the LRDG transported spies, counted enemy movements and supplies, and liberated POWs. As the RAF gained air supremacy, the LRDG shifted to raiding and sabotage, and disbanded in 1943, its members later participating in special operations in Italy, Greece and the Greek Islands, Albania, and Yugoslavia.

THE SPECIAL AIR SERVICE

Two other desert "private armies," the Special Air Service (SAS) Brigade and Popski's Private Army, operated independently and with the LRDG. The SAS was formed by a young major, David Stirling, of the Scots Guards and a Commando as well. The unit name masked its deep-penetration raiding and sabotage, and the fact that initially it was only about a platoon in size. Popski (an alias of Vladimir Peniakoff) operated more like the German Brandenburgers, sowing deception and confusion in enemy rear areas. The SAS suffered heavy casualties and disappointment when several major schemes miscarried, most notably at Tobruk, and its shift to a sustained combat mode was costly. Stirling was captured in 1943 in Tunisia. The small early raids of the SAS were far more effective, and less bloody, as they destroyed hundreds of German and Italian aircraft. The SAS label was later applied to an airborne brigade designed for behind-the-lines work in Europe, composed of one Belgian, two British, and two French battalions. SAS units were retained in the British and Belgian armies after the war.

These groups seemed buccaneer-like to friend and foe alike. Their multinational, multilingual personnel, their contempt for even 8th Army's casual uniform regulations and discipline, and their customized vehicles and esoteric skills were colorful offsets to British setbacks and disasters from April 1941, when Erwin Rommel arrived in North Africa, to the great victory of the 8th Army at El Alamein in November 1942. As in other theaters, when major forces were stagnated or failed, firefly "mobs for jobs" lit up an otherwise very dark night.

THE U.S. MARINE RAIDERS

East of Suez, there was less publicity, more miserable fighting conditions, and a more ferocious foe. The most visible elite forces in the Pacific early in the war were the United States Marine Corps' Raider battalions. While marine corps of other countries have been relatively small and highly specialized, e.g., the British Royal Marine Commandos and Soviet

Naval infantry, the U.S. Marines grew into a full field army in strength in World War II. The Marines' strong links to Congress and aggressive courting of the media helped them fend off several attempts to pare down or eliminate the Corps.[6]

Whether U.S. Marine training, especially brutal since the 1920s, in fact produced higher quality troops than first-rate Army units will long be argued. In any case, the Marines were the first U.S. ground forces committed to the offensive in the Second World War. The 1st Marine Division landed on Guadalcanal in early August 1942, and soon afterward, Marine pugnacity was demonstrated in a raid against Japanese-held Makin Island by the Marine Raiders, whose fearsome demeanor on Guadalcanal led Eleanor Roosevelt, the President's wife, then visiting the Southwest Pacific, to urge they be quarantined before returning to civilian life. In the Makin raid, dramatized and magnified in the film *Gung Ho*!, just over two hundred men of the 2nd Raider Battalion, "Carlson's Raiders," were transported to the Japanese-held atoll in two old large mine-laying submarines. While their impact was light, and losses higher than expected, Makin was an attack after many defeats and retreats. Unfortunately, it led the Japanese to upgrade their defenses, leading to heavy losses at Tarawa. At the time, however, like the raid in April by U.S. Army Air Force B-25 medium bombers launched from U.S. Navy carriers against Japan led by Jimmy Doolittle, it served as a propaganda tonic.

It was less than coincidence that President Franklin D. Roosevelt's son, James, a Marine reservist, was executive officer of the 2nd, commanded by Lieutenant Colonel Evans Carlson. He and his father, the President, had pushed hard, against much resistance from senior Marine officers, to have a commando-type force created under Navy control. Four Raider battalions were ultimately formed at their urging, although Marine regular officers resisted the idea of an elite within an elite, and argued that they were already in the raiding business.

The 1st Raider Battalion, headed by Lieutenant Colonel Merritt Edson, raided Tulagi and Savo Island in August-September 1942. Then, in the classic pattern, they were committed to sustained combat on Guadalcanal to the point of shuffling, mumbling exhaustion. On September 8, the 1st raided Japanese supply dumps and contaminated Japanese foodstocks. Soon afterward, four hundred Raiders of the 1st beat back an attack by 1,800 Japanese. From November 12 to December 4, 1942, "Carlson's Raiders" left the Guadalcanal perimeter in a 150-mile pursuit of the Japanese 228th Regiment, fighting a dozen actions and killing approximately five hundred Japanese at the cost of sixteen Raiders killed and eighteen wounded.

Until the Raiders were disbanded in January 1944, the Army and the Navy wrestled for control of the program. In June 1943, the 1st Raider Battalion fought alongside Army troops on New Georgia, later joined by the 4th Battalion. Casualties by

August had mounted to half the units' original strength, and neither saw action again.

The 2nd and 3rd Battalions went into action on Bougainville in the northern Solomons in September 1943 with the 3rd Marine Division, in a role indistinguishable from regular Marine units. In 1944, the Raider Battalions were incorporated into a new 4th Marine regiment to replace the unit lost in the Philippines, plans for a 5th and 6th Battalions were scratched and the Raider Battalion Training Center at Camp Pendleton closed.

Like the Marine Parachute Regiment, the Raiders' exact casualty figures were not published; like some airborne units in Europe, they were used in prolonged combat, although trained and equipped for short, intense actions; and like other "private armies," they were not popular with other units. Edson went on to win the Medal of Honor and retired as a major general. The more publicized and controversial Carlson had built his tactical ideas on concepts derived from observing the Communist 8th Route Army in China in the late 1930s. Wounded on Saipan in June 1944 and later promoted to brigadier general, he died in 1947.

THE "CHINDITS"

The Marine experiences, ironically, had little impact on the U.S. Army in the Pacific. Just as U.S. Rangers learned amphibious warfare and raiding from British tutors, the army's major long-range jungle raiding effort did not follow paths blazed by Edson or Carlson, but a trail hacked by the British innovator, Orde Wingate. His "Chindits" (a corruption of "Chinthe," the half-lion, half-eagle Burmese temple gargoyle, Wingate's symbol of air-ground cooperation), the Long Range Penetration Force, were raised in 1943 in India. Wingate, a zealous eccentric fresh from exotic exploits in Palestine and Ethiopia, had gained Churchill's support through the efforts of British military theorist, B.H. Liddell Hart. Insisting that only British and Gurkha troops and Royal Air Force ground teams be used, Wingate was concerned that some Indian units had performed poorly in Malaya and that the Japanese had raised an "Indian National Army" from prisoners-of-war. More practically speaking, the Chindits had to master English-language training quickly.

Results were uneven. In their long march into north-central Burma in 1943, supplied wholly from the air, eighteen percent of the 3,300 Chindits became combat casualties and twelve percent were laid low by disease. The Japanese felt little pressure from the raid. A refurbished and retrained Chindit force of six brigades--now called Special Force--landed in gliders behind Japanese lines in north Burma in the spring of 1944 and linked up with elements that had marched 400 miles.

Wingate's command was thrown into confusion by his death in an air accident on March 24, 1944. His successor did not know whether the Chindits were to cut Japanese communications supporting attacks on Kohima and Imphal, or to generally harass, or to spearhead an advance to the Irrawaddy. South East Asia Command, then headed by Admiral Mountbatten, put Special Force under American General Joseph Stilwell on May 16. Wingate had insisted that his men be kept in action only ninety days, but Stilwell's Chinese troops were blocked at Myitkyina in north Burma by a much smaller Japanese force in what Field Marshal Lord Slim called an "untidy, uninspired, ill-directed siege."[7]

Although he had rejected earlier British offers of help, "Vinegar Joe" kept Special Force along with "Merrill's Marauders" in action far beyond their limits, and by the time he heeded a direct warning from Mountbatten underlining his responsibility for keeping helpless men in combat, both units were reduced to human scrap. That fate, in fact, had been recognized in its original design. Galahad Force, like the First Special Service Force, was a child of the Quebec Conference, raised in late 1943, first as the 1688th Casual Detachment and then as the 5307th Composite Unit (Provisional). Volunteers were later resentful, after a ninety-day planning guideline became a promise, partly due to joking reference by the 5307th's commander, Brigadier General Frank Merrill, at a party in India. The unit was dubbed "Merrill's Marauders" by *Time* correspondent James Shepley. The men of the 5307th became disaffected after the unit was kept in combat far beyond three months,[8] never realizing that they had been deemed expendable from the start. Nor was Galahad an elite force in terms of selection and training.

Since jungle-experienced troops in MacArthur's command did not respond to the call for volunteers, most of the Southwest Pacific area quota was filled by men without battle experience. Many volunteers came from U.S. forces in tropical areas like Panama and the West Indies, and the overall profile was seen as substandard. Although U.S. Army Chief of Staff General George Marshall meant the force to be trained and fielded by Wingate, command politics in the China-Burma-India theater altered that part of the design.

The novelist John Masters, a senior officer in Special Force, painted the most grim and vivid picture in *The Road Past Mandalay* of the extended campaign's affect on British Empire troops. In the 5307th, conditions were perhaps worse because Galahad had no medical evacuation system. The men were so resigned to dysentery that one company cut the seats out of their pants.[9]

After minimum evacuation was ordered and hospitals culled, fifty of two hundred Marauders returned to duty were immediately sent back to the hospital by officers. After being pulled out of the line, the survivors, less than a third of the original force, went into a state of virtual mutiny. Just before it returned to the U.S., Stilwell entered the Marauders' rest camp, which military police refused to patrol. Courageous and alone, he faced men subsisting on local highproof, marijuana-laced "bull-fight brandy." No counterparts to the Chindits or Marauders were created in the British or American forces after the war, although the Special Forces' lineage is traced, rather strangely, to the 5307th and the FSSF.

THE "FLYING TIGERS"

Aerial warfare was also a fertile ground for creation of *corps d'élite*, beyond the fact that pilots have been seen as an elite caste since the dawn of the aerospace age. In the First World War, air forces were known for their chivalry, especially among fighter units, symbolizing individual leadership, style, sporting behavior, and freedom from immediate control of headquarters in action.

Beyond that overall mystique and the special aura around "aces" and survivors of many bombing missions, many air units were created as, or came to be seen as, elites per se (although not universally), e.g., RAF Bomber Command's Path Finder Force; 617 Squadron (the RAF's dam busters); U.S., British, and Japanese carrier pilots; the *Stalinfalken*; the *Luftwaffe's* Squadron of Experts (JV44); the Polish squadrons of the RAF; the "Abbeville Boys" (JG26); and *Nachtjäger Geschwader* 4, a night fighter unit which shot down 579 Allied bombers, losing only 450 air and ground casualties during the war--about a ten to one ratio.

Perhaps the best-known to date, the American Volunteer Group, popularly known as the "Flying Tigers," was an elite drawn from an elite and refined further by training and combat. The AVG's experiences paralleled some elite ground forces', in as much as courage and skill compensated for deficient equipment and the unit was disbanded after heavy losses and disaffection.

The AVG's popular image came from its Curtiss P-40B fighter aircraft's large ventral radiator scoop being painted with a tiger's snarling mouth, ostensibly to frighten Japanese pilots. Second-rate by European standards, the P-40B characterized Western aid to China--symbolic, obsolescent, and uneven. In the late 1930s, the Germans, Russians, and Italians had each helped the Nationalists at various times, as their forces engaged the bulk of the Japanese Army from 1937 to 1945. Not until late 1940 did the United States and Britain move to help the Chinese Nationalist forces in earnest.

A Chinese mission to the United States in 1941 was allowed to recruit for the AVG at U.S. Army and Navy bases, although it was illegal under U.S. law. Under lucrative one-year arrival-to-departure contracts, one hundred and one Army pilots sailed on a Dutch ship with a U.S. naval escort on June 9, 1941. Their commander, Claire L. Chennault, a former Army Air Corps officer, had been a victim of bomber-fighter faction-fights. As a mercenary pilot in China in 1937-39, he became a close friend of Chiang Kai-Shek.

The AVG was screened further upon arrival, and eighty-four pilots were trained secretly at British bases in Burma. Chennault's tactics pitted the Curtiss P-40B's strengths against the weaknesses of lighter, more maneuverable Japanese fighters.

They were tested in combat in Burma in December 1941, two weeks after America entered the war. In its seven-months existence, the AVG destroyed 299 Japanese planes confirmed, and 253 probables, losing twelve planes in the air, sixty-one on the ground, and twenty-six pilots--ten of them in accidents.

As the AVG was absorbed into the U.S. Army Air Force, things went awry. Chennault had tolerated much indiscipline, including one virtual mutiny. An overbearing Air Corps general and USAAF physical standards imposed on fatigued men led to only five AVG veterans receiving Army Air Force commissions. Others returned to the Navy and Marines or took lucrative jobs with Chinese and Indian air lines. Chennault went on to command the U.S. 14th Air Force in China, whose men wore a Flying Tiger patch.

THE BRANDENBURGERS

In spite of all their setbacks and frustrations, the Allied *corps d'élite* were powerful symbols, with impact beyond their numbers among friends and foes. In the European theater, when Hitler decided in 1943 to fight fire with fire, the so-called Brandenburgers were his point of departure, a unit formed in October 1939 by the chief of German intelligence, Admiral Wilhelm Canaris, for clandestine raids, deception, and infiltration.

Less visible than their Allied counterparts, they first saw action in the Polish campaigns in 1939 and later served in Africa, in the Balkans, and on the Eastern Front. One Brandenburger team landed in Southwest Africa from a submarine. Including recruits from "non-Aryan" ethnic groups, e.g., Indian army deserters and POWs, Afghans, and Balkan Muslims, in late 1944 the brigade-size organization was merged into the *SS Grossdeutschland* Division, and later formed into a division per se.

Brandenburgers became most visible when operating in small detachments under the command of Otto Skorzeny, an Austrian *SS* junior officer who caught Adolf Hitler's eye. He directed several hairbreadth operations, the most spectacular being the rescue of Benito Mussolini in September 1943 from a remote mountain resort at Gran Sasso in Italy guarded by pro-Allied Italian forces. Skorzeny, with a team of glider-borne paratroops and Brandenburgers, rushed the astonished guards in a relatively bloodless coup. Insisting over the pilot's objections that he fly out with Mussolini in a light plane which barely cleared the impinging peaks, Skorzeny became a matinee idol among the Nazi hierarchy, although some of the paratroops saw it as "grand-standing."

Soon afterward, again with Brandenburgers and paratroops, he kidnapped Admiral Miklós Horthy in Budapest to block a separate Hungarian peace. After he was promoted to *Sturmbannführer* and decorated by Hitler, Skorzeny's greatest tactical failure came during the Battle of the Bulge in December

1944. Paradoxically, it had the greatest strategic effect. His 150th Panzer Brigade included Brandenburger personnel and a special unit, the *Einheit Steilau*, which became the focus of Allied concern during the Ardennes battle. Some of its members spoke idiomatic English, and wore U.S. and British uniforms. Their mission was to infiltrate American lines and quickly seize the Meuse bridges, to help major German units reach fuel dumps near Liege and move on to split the British and Americans. A German order asking for volunteers and American vehicles and mentioning Skorzeny fell into U.S. hands on October 30, 1944, generating a 1st U.S. Army intelligence prediction of a German attack in December, although its guess on location was off the mark.

Some *Einheit Steilau* troops speculated about the "big picture," and it became common gossip that Skorzeny's mission was to kill Dwight D. Eisenhower. When the German attack began on December 16, paratroopers, not part of the 150th, were dropped along the front, raising Allied anxieties. Two *Einheit Steilau* teams were captured and later executed as spies. A German junior officer under interrogation related the barracks rumor about the mission to kill Eisenhower as though it were an official order. As a result, American troops around the Bulge, in near-panic over initial German success and confused by fog and snow, sought to thwart Skorzeny's "mission." Disruption far exceeded the teams' limited success in misdirecting units and performing sabotage. Eisenhower and other major commanders were immobilized under close guard during the battle's early phase. By December 20, as the 150th went into direct combat, Skorzeny was wounded in the head. Serving the rest of the war as a conventional unit commander and confined without trial by the Allies in 1945, Skorzeny escaped to Spain in 1948. He became a civil engineer and the focus of many Cold War rumors and suspicions that he was involved with "ODESSA" and *der Spinne*--the Web--a conspiracy of ex-Nazis.

THE AIRBORNE

Between the World Wars, the concept of parachute troops was developed most in the major nations defeated in the First World War, Russia and Germany. The Red Army was the first to drop airborne troops in combat, against Basmachi insurgents in Central Asia in 1928.[10] In the early 1930s, Deputy Commissar of War Mikhail Tukhachevsky created an independent airborne section of the Commissariat of Defense.

Views in the West ranged from awe to skepticism. Secret cooperation between the Weimar Republic's *Reichswehr* and the Red Army led to the creation of a German experimental airborne staff in 1935 headed by Brigadier General Kurt Student, and formation of Army and *Luftwaffe fallschirmjäger* battalions. At the same time, Josef Stalin's purges of the late 1930s weakened the Soviet airborne program. Soviet paratroops fared poorly in Finland during 1939-40; their drops were badly scattered and they failed to gain objectives. In the Second World War, most Soviet airborne troops fought as infantry or in small groups in guerrilla support, reconnaissance, sabotage, and intelligence roles. Two brigade-sized operations in 1943 had marginal success. After World War II, the Soviets reorganized paratroop forces under the Ministry of Defense, and by the 1960s reportedly had 100,000 men and enough airlift for 20,000. Their current strength is believed to be seven full divisions and an unknown number of smaller units.

The Germans first used airborne forces in the invasion of Denmark and Norway in April 1940, as small *fallschirmjäger* detachments seized critical points. In the onslaught on Belgium and Holland on May 10, 1940, German airborne forces reduced the modern, well-manned Belgian fort at Eben Emael, and seized airfields and bridges in Holland. Although a shock to the Allies, the *Luftwaffe* lost more than two-thirds of 450 troop carriers committed, and casualties were high among parachutists and glider troops.

The British Imperial General Staff feared a major parachute attack throughout the summer and early autumn of 1940. German invasion plans included a major drop in Kent by a paratroop division of over 7,000, followed by an air-landing division of 300 gliders and 1,000 transports.

The British held three mobile divisions in reserve to counter airborne attacks. An anti-parachute alert was conducted on September 7 when the aerial Battle of Britain climaxed and invasion was expected hourly, producing much confusion and tighter rules on giving alarms.

The threat never materialized. German airborne forces next struck in the Balkans on April 25, 1941, when two battalions of parachutists tried to seize the Corinth Canal bridge and block the Greek and British retreat, but the bridge was destroyed. In late May, Operation *Merkur*, a landing on Crete, began a full-scale and bloody operation. After German parachute and glider troops seized airfields on the narrow northern coastal plain, mountain troops were flown in to clear the island. British Empire and Greek forces on Crete totaled about 40,000, many fatigued from the evacuation and short of equipment. Of the 25,000 Germans, slightly less than half were airborne, the rest mountain troops.

Ultra decryptions of *Luftwaffe* command radio messages and the capture of German airborne manuals in France gave British and New Zealand defenders an edge, while German intelligence misestimated the defenders' strength and location. The victory cost almost three-quarters of the *Luftwaffe*'s Ju-52 transport planes and 6,580 lives. The American and British armies, unaware how severe German losses had been, hastened to form airborne divisions, while Hitler had told Student: "Crete proved that the days of the parachute troops are over. The parachute arm is one which relies entirely on surprise. In the meantime, the surprise factor has exhausted itself."[11]

When Student proposed a German-Italian airborne assault against the crucial British base at Malta in mid-1942, Hitler rejected the plan, apparently disdainful of Italian follow-up force. Malta continued to strangle Rommel's supply line. Another proposal by Student in July 1943 to drop two German divisions on Sicily was also turned down. One division was dropped three miles behind German lines as reinforcements, coinciding with a descent by British paratroopers and resulting in a free-for-all. A German battalion-sized drop on Leros in the Aegean in November 1943, along with amphibious attacks, captured 9,000 Allied troops. German parachute forces made no more large-scale assaults, but the airborne mystique was used to motivate young volunteers. Although by mid-1944 only a third of Germany's 150,000 paratroops had actual jump training, the Western Allied analysts rated paratroop divisions as the best in the German Army.[12] The last sizable operation came in the Battle of the Bulge in December 1944 in Operation *Greif* (Condor), as 500 *fallschirmjäger* were dropped haphazardly on both sides of the front. Some landed fifty kilometers outside drop zones, adding to panic caused by Skorzeny's operations.

Among the Western Allies, the British had led the way in 1940-43 in reaction to German success. The Inter-Service Training and Development Center which had studied airborne warfare in the 1930s was disbanded after the war began in 1939, but in early June 1940, Churchill proposed a 5,000-man parachute force. Transport aircraft were scarce, and the RAF reluctant to release pilots or planes from its first love, the bomber offensive. As the British Army rearmed and trained after the Dunkirk evacuation, many saw elite forces as a nuisance. Churchill sent prodding memos and visited airborne units. Their commander, Major General Sir Frederick "Boy" Browning, imposed Guards discipline on volunteers from many regiments to gain respectability in the eyes of powerful Regulars.

On February 10, 1941, after 38 British parachutists with Italian guides in Operation *Colossus* blew up the Tragino Aqueduct in southern Italy, their pick-up submarine failed to appear and they were captured. A year passed before the next British airborne action, Churchill furious--in the aftermath of Crete--that a whole year had been lost. In February 1942, a company dropped at night at Bruneval into France, escorting a radio expert who dismantled a German *Würzburg* radar set, then withdrew by sea. In November 1942, British airborne battalions dropped into North Africa. In the invasion of Sicily in mid-1943, many Anglo-American gliders crashed at sea. Scattered dropping forced a single British platoon to attack the Primosole bridge, originally a battalion objective. In the invasion of Italy in September, the 1st British Airborne Division landed at Taranto--from warships.

After ground fighting in Italy and drops in Normandy in June 1944 and in southern France in August, the British 1st Airborne Division spearheaded Operation *Market Garden* in September 1944, which was designed to seize three main bridges into northern Holland and then cross the Rhine tributaries and fan out into north Germany (but mainly to overrun German V-1 "buzz-bomb" launch sites in Holland). Field Marshal Bernard Montgomery later called the plan his greatest mistake.[13] The U.S. 101st and 82nd and the 1st British Airborne Divisions were dropped from south to north, respectively, at Eindhoven, Nijmegen, and Arnhem, seizing the key bridges for British 30 Corps' advance. The attack was a series of disasters. While American divisions were dropped all at once on the first day close to their objectives, the British landed farther away from theirs because planners wrongly believed soil around the key bridge was too soft. Due to aircraft shortages, they were committed on successive days. Worse yet, two SS panzer divisions were in reserve near Arnhem and General Kurt Student, founder of German airborne troops, was on the scene. Possible betrayal by a Dutch partisan was later overblown--treason could not have made things much worse. Bad weather hampered resupply and reinforcement, while 30 Corps moved sluggishly. The British 1st Airborne, with light weapons and no reserves and reinforced by the Polish Parachute Brigade, held out four times longer than expected, but less than a fifth of its original 10,000 men returned to Allied lines.

In April 1945, in the last major airborne operation, one British and two U.S. airborne divisions landed across the northern Rhine within range of supporting artillery on the west bank. Given the overall record of losses, the mixed results, and Britain's scarce resources, a major British airborne program was a luxury. Airborne divisions were often out-of-action; the 1st, for example, was in reserve from June through September 1944. RAF air transport was scarce and interservice relations poor. The airborne mystique, however, was a powerful force beyond cold numerical analysis. As Britain in 1940-42 faced bearing the brunt of the reinvasion of Europe, to Winston Churchill and many others, parachute troops paradoxically symbolized both a lost age of heroic combat and modernity. Laurence Olivier's film *Henry V* was dedicated to the commandos and airborne forces.

The large American airborne forces of World War II, like the British, were a reaction to German airborne operations in 1940-41. A small experimental platoon, formed in early 1940, in September became the 501st Parachute Battalion, and the German victory on Crete led to three more battalions and a Provisional Parachute Group being formed by September 1941.

The Army Air Corps, seeking to control airborne troops as the *Luftwaffe* did most of Germany's, held back on transports, bombers, and fighters, all in short supply. In the September 1941 maneuvers, errors and problems abounded. The Air Corps was not the only obstacle; in the tradition of the U.S. Army, the branch chiefs were uneven in support of airborne development. In March 1942, after the U.S. entered the war, an airborne command was formed. Lieutenant General Lesley McNair, Chief of Army Ground Forces, concerned about the growing number of specialized units, suggested breaking up the airborne divisions, keeping parachutists separate, and organizing the remainder as light divisions. But European theater planners wanted airborne divisions to equal standard divisions in strength. They did not, ultimately, but came closer than McNair would have preferred.

Under the 1942 War Department reorganization, an Airborne Center was established for training and doctrine, but airborne was denied branch status. In the long run, it made little difference. By 1945, five American airborne divisions and several independent airborne regiments had been formed. For the next twenty years, the airborne would upstage the formal branches. Drops involving one or more U.S. airborne divisions were made on Sicily, at Salerno, Normandy, southern France, Eindhoven-Nijmegen, and in crossing the Rhine. As with the German and British, the record was uneven. U.S. parachute forces were misdropped in North Africa and Sicily and fought for extended periods as regular infantry in Italy, Normandy, Holland, and at Bastogne, without heavy weapons. In Europe, many operations were cancelled, and the senior U.S. ground commander, General Omar N. Bradley, ordered proposed drop zones to be overrun to force diversion of planes to supply.[14]

In the Far East, American airborne operations were on a smaller scale than in Europe. On New Guinea, at Nadzab in September 1943 and at Noemfoor in July 1944, paratroops descended among friendly troops and were screened by smoke. Especially dramatic was the 503rd Parachute Infantry attack on the Corregidor fortress in 1944, a low-altitude jump which gained surprise--and disabled ten percent of the assault force. The 11th Airborne Division led the drive on Manila on the ground. One of its regiments jumped on Tagaytay Ridge, to no purpose. Only a third landed in the drop zone and half were slightly injured. A drop by a battalion of the 511th in the Cagayan Valley was equally fruitless. A great success, however, was the "triphibious" descent on the Los Baños prisoner of war camp. After the war ended, the 11th Airborne Division led U.S. forces into Japan as General Douglas MacArthur's bodyguard.

Lack of air transport and great distances limited airborne operations in the Pacific, but they were not used to advantage even when real opportunity existed in some island assaults. The quality of airborne troops as infantry outweighed their skill as parachutists in the eyes of commanders in both theaters. U.S. Marine parachute battalions (later a regiment) were never dropped in combat but used as conventional infantry on Gavutu, Tulagi, Choiseul, and Bougainville, suffering heavy losses in each case.

Japanese airborne forces, trained in late 1940 by German advisers, included an army brigade and a "Special Naval Landing Force." The Japanese conducted four large-scale parachute operations in the Pacific war: on January 11, 1942, in North Celebes in a diversion; on February 14, 1942, at Palembang on Java, suffering heavy losses, but preventing Allied oil field demolition and diverting forces from a nearby amphibious landing; on February 20, 1942, on Timor as part of an encircling force; and on December 6, 1944, near San Pablo in the Philippines. The first three succeeded, but the last dropped into the 11th U.S. Airborne Divisions area.

Airborne and air special operations in Burma were unique relative to other theaters of war. The air-supplied roving-column concept developed by Orde Wingate (discussed at greater length in the section on the Chindits) was designed to avoid pitched battle and force the Japanese to cover vulnerable points in the rear. The Chindits, supplied and sometimes delivered by air, relied on air power for artillery support, supply, and ambulance service. Weak Japanese air cover made the concept viable, but its effect is still debated.

In airborne operations in Burma, gliders dominated and paratroops were marginal. The Indian Army Parachute Brigade fought mainly in ground actions. The unorthodox, functional approach of Wingate and the head of the "Air Commandos" Colonel Philip Cochran (prototype for the character Flip Corkin in the *Terry and the Pirates* comic-strip) contrasted with other theaters of war. Their collaborative efforts culminated in the summer of 1944 in a glider landing of 14,000 men on a single jungle airstrip, who, with an overland column, operated for several months completely sustained and supported from the air.

However misused and misunderstood, airborne forces captured public imagination as a special elite based on an ordeal that tested courage and will before combat. Jump pay and special insignia and uniforms added to their mystique. Airborne forces produced a new image of combat leaders, their generals flat-stomached and young-looking. On the other hand, gliders, although successful and very dangerous, did not capture the military or public imagination. Although only no major combat drops were made after 1945, the airborne mystique continued to affect several parent armies. Earning the paratroop badge became mandatory for Israeli infantry officers, and the motto "The Brave to the Paratroops" symbolized status as Israel's ground forces' principal *corps d'élite*.

In the United States, the young airborne division commanders at war's end, James Gavin, Maxwell Taylor, Matthew Ridgway, and a host of their subordinates rose to senior command over the next generation. A sizable fraction of the U.S. Army was formed into parachute units, although airlift was scarce. American airborne units spearheaded several Cold War operations, e.g., Lebanon (1958) and the Dominican Republic (1965), and served as military police at Little Rock (1957), Oxford, Mississippi (1963), and the Selma March (1965). In 1956, airborne training became mandatory for American Regular Army lieutenants, the airborne mystique growing as its tactical utility shrank, amplified in such films as *Battleground* (1950), *The Man in the Grey Flannel Suit* (1956), *Ocean's 11* (1960), and *The Longest Day* (1963).

Ironically, the gaudiest flower of airborne forces bloomed in that twilight of obsolescence in France, a relative late-comer to the airborne arena. In May 1940, when Germany had 100,000 qualified airborne troops, France had only two companies of parachutists, sent to Algeria after the armistice.

For the next five years, French airborne units were raised and equipped by the British and Americans. Two battalions formed in England became part of the Special Air Service Brigade, performing special missions in France and with the British in North Africa. The 2nd jumped into France on June 5 and later fought in the Loire campaign and at Bastogne, the 3rd dropping in small groups among resistance forces and suffering heavy casualties. The 1st Assault Parachutist Battalion, formed at Fez in 1943 under French Air Force control and based on U.S. organization, never jumped, ending the war in reserve during the invasion of southern France. At the end of 1945, the French Army formed the 25th Airborne Division from the 25th Infantry Division.

In 1946, two paratroop battalions sent to Haiphong in Indochina became demi-brigades of Parachute Commandos, operating in Laos and Cochin-China. In 1947, a demi-brigade from France was converted into the 1st and 2nd Foreign Legion Parachute Battalions. Used extensively on raids and dozens of relatively small operations, French paratroops became the focus of world attention in the spring of 1954 at Dienbienphu. In the autumn of 1953, the French set up a blocking position in a bowl in the hills to threaten the Viet Minh insurgents' approach to Laos. Supply, reinforcement, and evacuation of wounded were to be by air. A cavalryman commanded the launched light armored attacks against surrounding Viet Minh forces and supply lines. The Viet Minh brought up Chinese-donated artillery quickly silenced the French guns, closed the airfield, knocked out the tanks, and closed in with trenches and mass attacks. Four days after the siege began, the airborne commander at Dienbienphu took over.

DienBienPhu was much like Arnhem--a badly conceived plan, confused command-and-control, lack of surprise, surprisingly heavy anti-aircraft fire, poor communications, supply failures, at the edge of support aircraft range, weak intelligence, and poor terrain choice. Supplies often landed outside French lines. When it ended, of seventeen parachute battalions sent to Indochina, eleven were captured at Dienbienphu.

Even as the survivors sat in Viet Minh "re-education camps," the first French airborne units arrived in Algeria. Eventually over two divisions of paras were committed, salted with Far East returnees after November 1955. While Indochina had not been popular, the Algerian war was widely detested in France. The French Communists, passive on Indochina, became actively hostile, and lack of support for the war was reflected in many reservists' and conscripts' weak performance. Tactics were revised; general forces made sweeps, seeking contact, and, when insurgents were contacted, paras arrived, usually in helicopters, for the close-in fighting.

When the war shifted from the countryside to the streets of Algiers, soldiers became political warfare specialists. Methods learned in Indochina--slogans, posters, and demonstrations--became weapons in an ideological war. Growing awareness of torture by French troops brought further public revulsion in France. A junta of airborne colonels, threatening to descend on Paris, ended the Fourth Republic which had betrayed them in Indochina and brought Charles de Gaulle to power, who ended the *paras* brief sway and took France out of Algeria. For a brief time, however, French airborne colonels had become media stars. One, Marcel Bigeard, proclaimed: "The men of the regiment are handsome, proud and courageous . . . we make real men out of them; healthy, sporting types, humane and well-bred,"[15] words reminiscent of Brigadier de Gaulle's vision of an "army of the future" in the 1930s.

Civilian political control of a military elite force in a Western democracy had failed dramatically. Gilles Perrault, an ex-*para*, viewed popular admiration fanned by the press as dangerous, pointing to parallels in values between *paras* and delinquent street gangs and the *SS*. Others called the *paras* Fascist, noting many German Foreign Legionnaires. The paras had cultivated a nihilistic posture, adopting Edith Piaf's song "Jé Ne Regrette Rien" ("I Regret Nothing") as she had adopted their mystique in "Mon Légionnaire." Their training stressed inferiority of civilians, and techniques of using *plastique* high explosives for terror and torture were taught.[16]

Political extremism and insubordination emanated most strongly from the parachute regiments and the Legion as the regular-reservist split became one of regulars vs. regulars. A "Secret Army Organization" was formed as a terrorist group that attacked the army and government in the aftermath of the failure of a putsch in 1961. Some paratroop units were disbanded, and 32,000 *paras* were transferred and given new, less dramatic uniforms and headgear. Colors and red berets were solemnly burned. The landscape was dotted with ironies. The Algerian war saw the helicopter upstage parachutes as the main form of rapid tactical aerial delivery of infantry. The Suez affair in 1956--another airborne operation whose success was nullified by politics--had added to the *paras'* frustration. The *parachutistes* going from avant-garde to anachronism in half a generation, and not only in France.

From 1955 on, French, American, and British airborne doctrine in the "nuclear age" moved away from large-scale tactical parachute operations. Interest grew in small units doing special jobs, using parachutes for surprise, like the Belgians at Stanleyville in 1964, or small boats, aqua-lungs, light aircraft, or vehicles. In 1968, Soviet paratroops disembarked from "civil" aircraft landing under false distress calls, opening the way for a Soviet invasion of Czecho-slovakia--the "counterrevolution." In the West, it became more and more apparent that even if airborne forces were dropped en masse and formed an effective, cohesive tactical unit quickly in a major war, the large arrays of planes required would be easy targets for a nuclear air-burst, massed anti-aircraft gun fire, and/or rockets.

By 1970, the helicopter upstaged dropping paratroops as a rapid and mass deployment technique in several countries. Nevertheless, the mystique endured. In the post-World War II British army, a disproportionately large number of citations went to the Parachute Regiment, although it usually went into action other than by parachute, as, for example, in the Falklands War. Paratroopers often played constabulary roles, e.g., in Greece, Palestine, the United States, Cyprus, the Dominican Republic, the Congo, and Ulster.

CYBERNETIC ELITES

A contrary line of development paralleling the proliferation of heroic *corps d'élite* was the appearance of mechanical and cybernetic elites. By the end of the First World War, technical skill and brainpower were clearly military resources of the first magnitude. Over the next generation, radar, napalm, mass-produced penicillin, the proximity fuse, the electronic digital computer, V-2 rockets, and jet fighters dramatized the marriage of Athena and Mars. New military elites appeared defined in terms of hardware and expertise. Cybernetic systems of humans and electro-mechanical devices went to war designed and controlled by scientists and engineers, with humans and machines critically interdependent.

Some, like crypto-analysis and electronic warfare units, were more purely cerebral, passively processing information rather than fighting. In others with direct combat roles, like the Path Finders, technical skills were foremost. Ironclad ships, tanks, and early aircraft were a step toward cybernation in the sense that they had extended human power through machinery. In World War II, however, "man-machine systems" became vital in waging war. Because some were secret and/or esoteric, they gained little recognition or glamour. It was neither easy nor wise, for example, to explain a photogrammetry machine in the Sunday supplements of 1942. But growth of these systems, sometimes in the face of high-level resistance, was proof of their utility.

THE PATH FINDER FORCE

When a special study in 1941 showed RAF bombing accuracy to be much poorer than had been estimated, the Air Ministry insisted a target-making force be created. Air Marshal Sir Arthur Harris, Chief of RAF Bomber Command, resisted, delayed, and deceived as long as possible, arguing a *corps d'élite* would drain Bomber Command of vital crews and sap morale.

Such a unit, moreover, would be a virtual admission that earlier claims were false. The unit's title suggested a navigation problem rather one of bombing accuracy, when, in fact, both were involved. The PFF was commanded by D.C.T. Bennett, eventually an Air Vice Marshal, a blunt Australian ex-RAF regular whose calls for a more realistic approach to bombing accuracy put him in line for command when the PFF was formed. Harris, however, missed no chance to reduce the force or minimize its worth.

Four squadrons with relatively good records were weeded out during a shakedown period, and the PFF went into action in August 1942. In twenty-six target-marking missions, the PFF missed six times in bad weather, but otherwise marked targets seventy-five percent of the time. Even though it led the way, its losses from August to December 1942 were two percent compared to thirteen percent for standard squadrons. The PFF filled the gap until electronic guidance systems for individual aircraft came on line in 1943 and 1944. (The Royal Navy Fleet Air Arm pioneered target-marking with lean resources in North Africa, 1941-42, in cooperation with Desert Air Force, navigation over trackless sands being much like flying over the ocean. Some of their methods were adopted by the PFF.)

The Path Finders did not depend on human navigators' skills, but used the Oboe system, guiding planes equipped with receivers to targets and signalling when to release bombs. A follow-on system, H2S, displayed a radar-scope image of the ground. The PFF was innovative in several ways. The unit developed pyrotechnic markers and Bennett was the first senior RAF officer to endorse FIDO anti-fog burner systems. The Path Finders also adopted the all-plywood Mosquito bomber, initially rejected by RAF group commanders. At the end of the war, the PFF was disbanded, Bennett went on to airline management, road racing, and politics. The PFF's highest tribute came posthumously from Dr. Joseph Goebbels, Nazi propaganda chief: "The English aimed so accurately," he noted in his diary on November 26, 1943, "that one might think that spies had pointed the way."[17]

"PHANTOM"

Since the Napoleonic Wars, the expansion and dispersion of armies and growing complexity have made information harder and harder to get to headquarters, even when communications techniques advanced. In all the major modern wars, commanders often lacked vital data about battles which they officially "commanded." The British Army sought to cope with this problem in the Second World War with the GHQ Liaison Regiment. Code-named "Phantom," its squadrons roamed the battlefield in light vehicles, transmitting information directly to Corps and higher headquarters, bypassing combat units' clogged and confused signal nets.

"Phantom" first served as a feedback loop to the British Army in France in 1940. After growing pains in the desert, 1940-42, it became a prime combat information source for major Allied commanders in Europe. While security kept the force out of public view, the Americans, dependent on the British at first, trained their own units, although they were not ready in time to take the field during the war. (The film star David Niven served with a "Phantom" unit for three years.)

THE 79TH ARMOURED DIVISION

The British Army's 79th Armoured Division was a true technical elite. While the men were a cross-section of the parent force, the 79th's machines were elite. Its "Rube Goldberg"-"Heath Robinson" fighting vehicles, known as "funnies," landed in segments with assault forces all along the British-Canadian beachhead on D-Day (June 6, 1944), rather than as a cohesive unit. Formed in 1943 from several specialized armor units, the 79th was commanded by Major General P.C.S. Hobart, a British armor pioneer recalled from retirement.

The 79th's "funnies" included amphibious "Buffalo" troop carriers; "Crocodile" flamethrower tanks; DD "Duplex Drive" amphibious tanks; "Crab" tanks, whose motor-driven flail chains beat a path through mine fields; Bobbin wire-screen "carpet-laying" tanks which created a more solid footing for wheeled vehicles on sand; bridging tanks; "Kangaroo" armored personnel carriers, made by removing turrets from obsolescent tanks; tanks for placing fascines; AVREs (Assault Vehicle Royal Engineers), mounting a short-range twelve-inch mortar--the "Flying Dustbin"--for shattering fortifications and was equipped with fittings for placing heavy explosive charges. U.S. forces used only "DD" tanks from the 79th's array of "funnies," and many American combat engineers and infantrymen fell on Omaha Beach trying to do the job by hand.

Later, the custom-built, flexible 79th performed various operations under the control of GHQ, such as besieging the Channel ports, clearing the Scheldt estuary, and crossing the Rhine. The division was disbanded after World War II, its only vestige a Specialized Armour Development Establishment, the concept of "funnies" virtually forgotten as the armies of the world relied on arrays of standard tanks, self-propelled guns, and personnel carriers.

UNDERWATER WARRIORS

An attempt to extend individual combat into the long inaccessible realm of the sea saw creation of perhaps the most rarefied elite forces to date, the frogmen--a fusion of human and life-supporting devices producing a bio-mechanical creature. While clandestine swimmers were hardly new to warfare, in the twentieth century new technologies of underwater breathing and propulsion allowed freer and longer movement, revolutionizing the tactics of underwater warfare. In the First World War, Italian junior officers destroyed three major Austrian warships with motorboats and torpedoes aimed by swimmers. While these efforts were not maintained, adventurous junior Italian naval officers revitalized the program in the early 1930s. In 1935, when Italy's invasion of Ethiopia brought on a crisis, the 10th MAS (*Motoscafi Antisubmergilio*) was formed, its motorboat flotilla label covering an experimental force which developed tactics and technology for attacking harbors with small submarines and swimmers using underwater breathing gear. The easing of war fears put the unit on the shelf, but, when Italy entered the war in 1940, the reconstituted 10th operated on the North African coast and around Algiers, its most spectacular successes the sinking of two British battleships at their moorings in Alexandria harbor, and a number of Allied cargo ships at Gibraltar using magnetized "limpet" mines. After the 1943 armistice, MAS veterans passed much information to the British and fought for the Allies in the northern Adriatic.

British frogman operations began as a response to the Italian challenge, using veteran salvage divers. Eventually, one- and two-man torpedoes undertook special operations, including an ineffective sortie against the German battleship *Tirpitz* in Norway. The Germans also used frogmen, including attacks against Allied shipping at Anzio and against bridges on the Rhine with limited success. American underwater warfare efforts were mainly directed at pre-invasion beach reconnaissance and demolition. In spite of a relatively slow start and primitive equipment, U.S. Navy Underwater Demolition teams preceded many landings in Europe, with heavy casualties, and in the Pacific, with much lower losses. The first UDT teams deployed from motorboats, wearing only goggles, trunks, and rubber fins, carrying special tablets and styluses to record obstacle data. Shore bombardments distracted the enemy, since UDT teams often operated in daylight within sight of friends and enemies.

The use of SCUBA (self-contained underwater breathing apparatus) was more prevalent in Italian, British, and German forces in World War II than in the U.S. Navy. But even that gear, which recycled oxygen for chemical purification to avoid tell-tale bubbles, limited diving depth to about 30 feet. The Japanese also used special swimmer units with minimal evident effect.

Further advances in underwater technology came to the public eye in 1956 when Commander Edward Crabb, a pioneer British frogman who had won the Victoria Cross in the Second World War, disappeared. Rumors circulated that he was killed or abducted while checking the hull of the Soviet cruiser *Ordzhonikidze* visiting Britain. Others suggested he had defected, having been passed over for promotion. Official British comment was vague, and speculations abounded.

Increased interest in oceanic exploration was stimulated by a French naval officer, Jacques Cousteau, who had developed the diluter-demand, compressed-air "aqualung," thus increasing the depth of free-diving tenfold. The search for undersea resources, the development of the "oceanic deterrent," and expansion in world submarine fleets stimulated interest and research in these areas, leading to reported experiments with membranes to allow direct breathing of sea water, special gas mixes for deep diving,and the use of porpoises.

IMAGES OF ELITES

Beyond their physical effect in operations and their role in diverting enemy attention and resources, *corps d'élite* also served a propaganda function, offering subjects for film, the press, and literature. In an age when the battlefield was bleak and uneventful much of the time, and combat rarely captured with any pictorial precision, images of paratroopers, mountain troops, and commandos dramatized war. In aerial combat, gun-cameras caught glimpses of action, jerky, brief, and fragmented, but photographs of direct actual ground and naval combat were extremely scarce. Only one picture of a Japanese surface ship was taken in combat from a U.S. Navy surface ship in World War II, and a very fuzzy one at that. In that vacuum of imagery, scenes of training and mock combat often substituted for real action as propaganda became a major weapon of war.

In a world of manipulated ideas, half-truths, and lies, elite forces of various nations came to represent values as well as functions. The Greek Sacred Company, *Hierchos Lochos*, for example, composed completely of officers of the Greek Army, embodied Royalism among Greek forces fighting with the British in the Mediterranean during 1941-44. Like British commandos, they were seen as inspiring heroes to some, and villains to others, an effect that would spread over time.

Especially paradoxical were Communist elite forces. Since Fascists were social elitists, their use of special units conformed to their dogma. But although ideologically Communists were egalitarians and levellers, in Russia, China, and Spain leaders of proletarian armies created elite forces.

After the Russian Civil War of 1918-22, a debate between elitists and populists in the Red Army command led to a compromise mix of regulars and territorial militia. During its first twenty years, the Red Army rejected such forms of the Tsarist system as officer's epaulets, saluting, formal ceremonies, modes of address, and a unilateral command system. The pendulum swung back slightly after the Finnish fiasco of 1939-40 when commissars were temporarily denied sharing power with unit commanders. The massive defeats of 1941-42 brought a series of changes, however, as the struggle against the Nazis became the Great Fatherland War--a patriotic defense of Russian soil. Party generals were eased out, and a much reduced authority awarded commissars, who adopted army uniforms. Nationalism overshadowed socialist ideas. Saluting, officer privileges, and tight discipline returned, and eventually a third of the Red Army units received "shock" or "Guards" titles. Special NKVD units, the Cossack Division, airborne units, Red Falcon fighter squadrons, and naval infantry were clearly *corps d'élite*.

After the 6,000-mile Long March, the Chinese Communists adopted a military discipline not unlike Cromwell's New Model Army, awarding elite status after the test of battle. The highest title in the People's Liberation Army, "Iron and Steel Brigade," was given to division commanders before an impending battle, as their divisional sub-units were "allowed" to compete for the coveted designation. In the 1960s, the PLA established "model" or demonstration units in the air force[18] and "Four-Excellence Companies" in the army.[19] Until the late 1950, at least, the Chinese Communist militia had been an "elite . . . a carefully picked minority."[20] This and the fact that the 1949 victory came after "units of the Kuomintang Army defected to lay the basis for the Red Army"[21] reflect a series of paradoxes.

THE WAFFEN SS

The most famous and lurid political-ideological *corps d'élite* was the *Schutzstaffel*--the *SS*. Heinrich Himmler, its founder and later head of the "*SS* state," saw the *Sturmabteilung*, the *SA* or Brown Shirt street army of the Nazis, as "infantry of the line" and the *SS* as the "imperial Guard of the New Germany."[22] Created in the mid-1920s, the *SS* grew slowly. A portion of post-World War I volunteer force (*Freikorps*) veterans, elite by their own definition, joined in the late 1920s, and some of their insignia and style were adopted by the Nazis. After 1929, middle-class professionals thrown out of work by the Depression flocked to the *SS*. When homosexuality among *SA* leaders led to public trials during election battles in the early 1930s, Himmler eliminated such members from the *SS*. In the "Night of the Long Knives" putsch in 1934, Hitler used the *SS* as the executioners, buying the army's favor by purging the *SA*--including Rhöm and his entourage and some innocent bystanders. The *Allgemeine* (general) *SS* expanded in the late 1930s into large formations. Its trappings and ceremonies shaped by Himmler's romanticist impulses. Although aristocrats appeared in higher ranks and the *SS* cavalry, most left after the putsch. Nazi "eugenic" selection standards were imposed, *SS* cadet schools established, along with fumbling experiments with economic autarchy and selective breeding, and the Burgundy scheme for creating a model National Socialist enclave.

The war saw an expansion of the *Allgemeine SS* into the *Waffen* (armed) *SS* in 1940, in which Orientals, Muslims, and Slavs joined Nordic elements in a vast multinational army; the image of elitism was more image than substance. It has been suggested that of the 600,000-men, thirty-eight-division *Waffen SS*, only the 18th *SS* Panzer Division (*Horst Wessel*) maintained a level of performance as a really elite field force.[23] Many recognized that *Waffen SS* enlisted men were of higher quality than the officers.

Slaughter of unarmed prisoners and civilians by some junior officers stained the *Waffen SS* image, as did the massacres of French civilians at Oradour-sur-Glâne and of American prisoners-of-war at Malmedy. Some Army officers suggested political selection of officers led to a squandering of valuable human resources. Rommel discouraged his son from joining the *Waffen SS* in 1943, deeming Heinrich Himmler as too unsavory a commanding officer.[24] The *Waffen SS* served as a "fire brigade," emergency reserve force in the last years of the Third Reich. By the end of the war, the seventeen all-German *Waffen SS* divisions had come to closely resemble Army units in performance and, in some cases, military customs and procedures. As its high losses on the Eastern Front became common knowledge, recruiting drives met resistance in Germany.

The *Waffen SS*'s sinister glamour has lived on, stirring admiration among some and revulsion among many, as dramatized at Bitburg in 1985. Like Napoléon Bonaparte's and Robert E. Lee's defeated armies, the *SS* became popular among military buffs and collectors. Even former foes pointed to distinctions between the *Einsatzkommando* and concentration camp guards and the relatively clean record of combat units. *Waffen SS* gained veterans' pensions from the Bonn government in 1961. Some marched on in fact as well as image as *Waffen SS* veterans saw combat in Indochina and Algeria as Foreign Légionnaires.

The record of the *SS* presented at Nürnberg and historians' analyses revealed a full spectrum of behavior. Some *Waffen SS* units adhered to the laws of war and engaged in acts of individual humanity. Others exulted in nihilistic danger-seeking or carried out spontaneous massacres. In a perverse irony, the *SS* upstaged the French Foreign Legion as the stock popular image of military elitism, its fame outlasting many of the forces that brought it down.

THE KAMIKAZES

The image and consciousness of death is a strong element in the elite force ethos. Desperate sieges, hopeless heroism, and last stands abound in their chronicles. Units of the French Foreign Legion, the Imperial Guard, and the Swiss Guards of the Bourbon kings proved dedication by fighting to the literal last man. Defenders of Masada, the Alamo, Chapultapec, and Camerone are extolled by Israelis, Texans, Mexicans, and Frenchmen, respectively. In any case, be it glory, fool-hardiness, or psychopathology, *corps d'élite* have a special contract with death. In that sense, the *kamikazes* of World War II were not just a bizarre flower in the Japanese garden, but the most extreme variant of a universal species.

Suicide is an old, strong theme in Japanese culture, e.g., romantic tales of young lovers hurling themselves into volcanos, ceremonial suicide (*suppuku*), and *hara-kiri*, the samurai's grisly act of atonement for dishonor, were popularly known in the West before World War II. From the beginning of the Pacific War, the Western Allies were amazed by frenzied *banzai* charges, tenacious defense of position, and refusal to surrender in hopeless situations. There had been many suicide attacks by Japanese soldiers and airmen, such as crashing a crippled plane onto an Allied ship or other military target, and in one instance diving into an American torpedo about to hit a Japanese carrier.

In 1944, after suffering a heavy defeat in the Battle of the Philippine Sea, the Japanese formed a "Special Attack Corps" to make up for the widening gap between U.S. and Japanese aircraft production and pilot training. Based on the cold logic that landing was the hardest part of flight training, they sought volunteers to crash explosive-laden planes into U.S. ships. The program soon became known as *Kamikaze* (Divine Wind), referring to the great typhoon that destroyed an invasion fleet of the great Kublai Khan sailing to conquer Japan.

At first, scattered *kamikaze* attacks in the Philippines in late 1944 did not seem different to Americans from the usual spontaneous suicide attacks. In the spring of 1945, however, U.S. Army and Marine forces landed on Okinawa, within aircraft range of the Japanese home islands, and the supporting naval forces were beset by a storm of blood and fire. American and British technological advantage denied victory to the grisly new force. Proximity fuses, which detonated near-misses of Allied anti-aircraft shells by radio signals reflected from targeted planes, had just come into use. Outlying radar picket ships allowed U.S. planes to intercept many *kamikazes* on their way and gun crews time to prepare for the bombs attached to the lightly constructed Japanese aircraft, which could not explode below ships' waterlines as naval shells could and thus penetrate armor before exploding. British carriers hit by *kamikazes* were relatively unaffected since they had armored decks. American carriers had lighter wooden decks, so that they could carry more planes.

A Japanese attempt to increase speed, explosive power, and penetration, the bomber-transported, rocket-powered *Oka* bomb, failed. Since *kamikazes* were an attempt to make up for poor pilot skills, the high-speed *Oka* made little sense, Americans calling it *Baka*, or foolish.

Nevertheless, *kamikazes* played the main role in sinking fourteen U.S. ships and damaging 172, including several heavy units, and in killing almost 5,000 American seamen. Although withdrawal was discussed and senior commanders were relieved to reduce stress, the fleet held on, and Okinawa fell in July. The 4,364 *kamikazes* gave better than they got numerically. Most, but not all, were volunteers.

THE BATTLEFIELDS OF TOMORROW

A variety of *corps d'élite* created before World War II were expanded, refined, or disbanded during the Second World War and in the many "small wars" that followed, such as mountain troops. The French *chasseurs alpin*, German and Austrian *gebirgsjäger*, and Italian *alpini* had all been formed earlier. The American 10th Mountain Division was the sole formal U.S. entry into this exotic realm, while there were variants in Warsaw Pact countries, Switzerland, and in India, Pakistan, and China. As the bibliography suggests, the French Foreign Legion, formed in the 1830s, has experienced swells of growth and shrinkage since 1939, as has its conceptual descendant, the Spanish Foreign Legion. The British Household Division has maintained its elite caché rather more than the U.S. Marine Corps which took in substantial numbers of draftees in Korea and in Vietnam retaining elitism mainly within such nodes as the Marine airwings and Force Reconnaissance elements. While this bibliography includes entries on such forces, the main focus is on the more recent *corps d'élite* and their links to "special ops."

The lists of special operations and historical and contemporary special units in the appendices underline how they have widened in the spectrum of world conflict in the nuclear age. Aside from their role as a rapid-response anti-terrorist force and in low-intensity conflict, a major potential role for elite forces is fighting on after a nuclear exchange, where, as in a full-scale conventional battle, their unit cohesion, skill, and dedication would be at a premium. Scenarios for prepositioning weapons and supplies, scavenger systems for "living off the land" in a heavily urban battle zone, and forms of transport not based on direct lines of supply have been proposed. Yet elite forces on one or both sides might only be spectators to a main battle in a conventional or nuclear theater of war, depending on whether the enemy commands the air, if weather would interfere with reinforcement, or if transport aircraft were preempted for supply or evacuation.

While allocating alternate roles for elite forces for critical-point security as a reserve or supplementary "palace guard" at home would offset such an implicit waste of special resources somewhat, that would fall far short of the rationale behind forming them. On the other hand, scenarios for their use well behind enemy lines in major conflict offer little more than an uncertain chance of survival.

Evidence regarding Soviet plans and intentions in an invasion of Europe is limited. Estimates range from massive mechanized thrusts, carpeting the battlefield with artillery and tank assaults, to a more deft use of bluster and mobilization to generate political pressure in Western Europe and block or hamper a coherent NATO military response. Images of refugees jamming the roads of Belgium and France in the *blitzkrieg* of 1940 and in South Vietnam in 1975 come to mind. Soviet *spetsnaz* units for deep penetration against critical points have generated much concern and speculation. One possibility is a battle for crucial communication-transportation nodes in Benelux-West Germany, involving small *spetsnaz* detachments, perhaps aided by sleeper agents and activists, stirring up a mélée in which nuclear weapons and unconventional high firepower military hardware could not be brought to bear. At Prague in 1968 and at Kabul in 1979, skillful *spetsnaz* air-landing operations suggested a degree of deftness well beyond the stereotype of a lumbering Russian bear.

Both a diffuse offensive using special units and a high-tempo armor offensive would put a heavy burden on Western defense and on small detachments of troops out of touch with neighbor units or headquarters with the temerity and skill to engage superior forces. Either situation would require tenacity tempered with flexibility, characteristics not usually associated with standard forces trained for linear warfare controlled by a "chain of command." It has been much argued in Western defense circles whether such difficulties can be offset by granting independence of action to units la the "optional command" tactics of the Israeli Army or the Bundeswehr's *aufträgstaktik*--general mission type orders. Such a system presumes very high levels of training, skill, and determination.

At the same time that numerical inferiority has demanded a qualitative offset, development of simple and cheap antitank and antiaircraft missiles has lowered the level of technical skill and manpower required to have large numbers of devices in place with a high first-shot kill probability. The demonstrated power of such new weapons in defense in the 1973 war shifted the Israeli Army's balance of forces away from primary reliance on elite infantry shock troops and armor to larger standard infantry forces and artillery support in order to protect tanks seeking a breakthrough point, and led to reconsiderations of force balance in other nations as well. Dispersal of security forces across a widely extended battlefield to provide a deep and broad defensive screen and increased need for "tube" artillery precision-guided munitions and for mobile anti-tank and anti-aircraft weapons in the battle zone all stem from the 1973 war. The new conditions suggested to some that small bands of missile controllers could be dispersed over the battlefield, with an implicit greater reliance on high-quality volunteer forces, further expanding the typology of *corps d'élite*.

However attractive to military and civilian policymakers, elite forces have also threatened free institutions. Like any "weapon system," they can be aimed in any direction. Thinking-out their role and side-effects has not always been kept in view in the press of crisis, nor has there been full awareness on the part of their civilian masters in calmer conditions. Whatever the apparent advantages, *corps d'élite* may generate political complexities far beyond those arising from the use of standard forces.

Advances in military technology have also shaped the nature of revolution and insurgency, a fact noted by such disparate observers as Edward Luttwak and Jacques Ellul,[25] both of whom noted how the increasingly delicate networks of high-technology societies has made them vulnerable to *coups d'état* and guerrilla warfare.

The image of tiny bands seizing or cutting vital nodes and links of urban society has in turn led to the raising of paramilitary forces--politically conscious, far from traditional concepts of military honor and international law, aggressive, skilled and sometimes thoughtfully cruel, as antidotes to the venom of terrorism. While much analysis of terrorism has focused on the search for patterns among terrorists themselves,[26] varieties of motive, style, and individual characteristics have caused frustration. As J. Bowyer Bell has noted:

> The profile of the terrorist is not possible given the present state of analysis. . . . [It] certainly does not permit the erection of elaborate theories or the construction of elegant models.[27]

Erosion of the soldier's code in limited war and resort by Western politicians to extra-legal means have moved in tandem. There are other basic congruities. Terrorists, military elites, and police special units all depend on small group cohesion, sense alienation from society at large, and often subordinate means to ends. Bell has noted how attributes of guerrilla terrorists "are not remarkably different from those of a conventional professional soldier."[28] Such data raises the question of how often, in the domain of political extremism, psychopathy is at work on both sides.[29] The central dilemma here lies in the balance between inaction and over-reaction.

A fundamental shared attitude of guerrillas, terrorists, and elite forces is their disdain of boundaries, ethical, physical, and behavioral. The desire of Secretary of Defense Robert McNamara to close the borders of South Vietnam in the 2nd Indochina War, however bizarre or grandiose, was based upon recognition of the fact that isolating insurgency is fundamental in bringing it to bay. The clear zonal frontier limits of Algeria gave French Army *paras* something definitive to chew on beyond the reach of the Americans in Vietnam.

Modern terrorism often leads regimes to rely on elite forces and covert operations that also defy or ignore limits. Thus, failures of commando raids may be more a strategic than a tactical failure, as was the abortive raid to free U.S. hostages in Iran in 1980. With political stakes out of proportion to numbers involved, combat between irregulars and security forces has become more and more cerebral and powerfully symbolic. The successes gained by the patient, long-haul, low-key style of General Sir Gerald Templer and Edward Lansdale reflects how a bluff, hearty lunge into stand-up battles can be inappropriate in "small wars."

While terrorist-guerrilla forces are affected by irrational and contradictory internal forces, so are elite forces. The image that rapier-like, intrepid, highly-trained troops offers to policymakers may not hold up when they go into action. Indeed, such forces' presence and methods can confound policy. The capture of Special Air Service (SAS) men on the Ulster-Ireland border in 1976, embarrassing links between former SAS men and a purported assassination plot against Togo's president in late 1977, involvement of former special operations personnel in mercenary forces in the 1960s and 1970s and in the Iran-Contra affair and the Philippines coup of 1987 suggest how elite forces often operate in a shadowy realm between law enforcement, the military, and clandestine operations.

In the cold war, elite units have often been used as the visible "cutting edge" in "brushfire" wars, as well as in constabulary and covert roles. Of course, when they are conspicuous, they can be used to mask other activities. Elite force leaders--in the West at least--have often been found at the extremes of the political spectrum, as have the symbols thereof. The role of the *paras* in bringing de Gaulle to power, John Kennedy's and Ronald Reagan's revitalization of special operations, and the dependency of Lyndon Johnson and Richard Nixon on symbols of military power in their hours of travail all reflect the political aspect of *corps d'élite*.

The triumph of highly-trained elite forces against terrorists in many instances since the early 1970s is not a guarantee of certain or indefinite success. Terrorist activity may ultimately produce greater international solidarity among security forces and cooperation to prevent such debacles as those battles in Uganda and on Cyprus between outside rescuers and local security forces. At the same tiem, the unheralded arrival of one nation's armed forces on the soil of another without previous agreement raises the question of whether uninational military elite forces are the best way to combat terrorism. Police-type forces, for example, recruited and supported on a retainer by a consortium of nations in exchange for reduced insurance rates for participants might have greater mobility and less symbolic and diplomatic vulnerability. While United Nations forces are anathema to many nations, a "private" non-profit organization might eliminate some problems of uninational deployment, including failure by a single country's counterterrorist forces. As with an antiterrorism treaty or an international terrorism tribunal, anything like such a force seems unlikely for the foreseeable future.

The escalating dialogue of cruelty seen in low-intensity conflict stems from insurgents trying to appear more than they are in the spirit of "Kill one; frighten a thousand." Uncertainty regarding the real power of terrorists and guerrillas is a basic dilemma in estimating and countering terror. Are three, three hundred, or three thousand behind simultaneous explosion of a dozen bombs or the massacre of a village? Visibility is the prime goal of terrorist activity, and a good part of the terrorist's game is intimidation and seeking recognition through reaction on the part of formal authority. Formation or deployment of elite forces as a response in recognizing a level of threat can be a victory in itself. Police forces were created in the early nineteenth century to give the state a response to threat short of committing troops. Terrorists stress the punitive and violent nature of the state they oppose, and acts of terror often seek to force the regime's repressive nature into the light of day.

Resort to military forms of force is in itself a statement about the seriousness of the terrorist threat. Many antiterrorist forces have been raised from police rather than military forces, while in other instances, military forces have been seen as relatively neutral versus police who are seen as brutal, corrupt, and/or inept.

In respect to elite units, creating organizations to perform tasks pitches them toward seeking out tasks which they were created to perform. To the extent that organizations develop habits and methods, those may be used against them. When they succeed, it may be spectacular; when they fail, it maybe more so.

Most elite forces have been created reactively, e.g., the proposal of Prime Minister Harold Wilson to create special antiterrorist squads in 1972 and President Jimmy Carter's "Project Blue Light" and Delta Force units based on cadres of U.S. Army Special Forces to counter terrorism and skyjacking. While such organizations serve as a cautionary symbol, they also formally validate the threat itself. That may weigh little against elite forces' forestalling of terror by sending a signal that stakes are very high and retribution likely.

Elite forces also proclaim confidence in the ability of individual courage, skill, and aggressiveness to offset both terrorism and mass armies' numbers and technologies. Reliance on their symbolic power runs counter to the theorem that "armies do not win wars by means of . . . super-soldiers but by the average quality of their standard units."[30] The environment of a battlefield dotted with elite force precision-guided missile controllers and quick-reaction forces to cope with mass armored attacks, and that of a terrorist emergency are quite different in essence and detail. Nevertheless, the scenarios of "New Battlefields," surrogate wars, and terrorism have stimulated the growth of these diverse flowers. For all their strangeness and beauty, they have often exhausted or poisoned the soil in which they grew and shown less hardiness than their gardeners expected. Historians of the future may determine how much the recent proliferation of these exotic species has been a product of policy and a coherent search for responsive instruments, of emotional reflexes, or of a trend toward the cybernation of war.

Those who shape, maintain, and control such forces may have to quickly sort through conflicting and uncertain data in crisis and depend on their intuition in deciding if or how they will be used. The political aspect is especially powerful at a time when military service has been widely rejected by middle class youth in the United States, and elite forces are a form of deterrence and symbolic aggressiveness. Usually, political support for use of military force is proportional to perceived distance, severity of threat to the society, and duration of operations.

Several elite forces created in the twentieth century had a strong political-ideological at the outset, such as the Spanish Foreign Legion, the International Brigades, the *Waffen SS*, U.S.

Army Special Forces, and the Red Guards. Some leaders of functional elite forces later played political roles, as did Karl Dönitz of the German submarine forces; Rommel, a *gebirgsjäger* officer in World War I and later commander of Hitler's bodyguard; Edwin Walker of the 1st U.S.-Canadian Special Service Brigade; David Stirling of the SAS; and Curtis LeMay of the Strategic Air Command. Some *corps d'élite* became cadres of or models for political parties or movements, or security forces in a political milieu; examples include the *arditi*, the *paras*, the *Freikorps*, and various elites in coups and revolutions in the Third World. In a parallel vein, Special Forces were patterned after various guerrilla cadres in the Office of Strategic Services (OSS) and the Special Operations Executive (SOE) in World War II. The re-formation of such units in the early 1950s fit the "captive nations" strategy which foundered in 1956.

THE COSTS OF GLORY: CORPS D'ÉLITE LOSSES

In the First World War, a grim pattern emerged in respect to *corps d'élite*, and the tradition carried on through World War II. Refined methods of picking the best of volunteers was paralleled by refined military technologies which increased the slaughter among those chosen, time and time again. A third of the German parachute forces died on Crete; three-quarters of the British 1st Airborne Division was lost at Arnhem; half of the British Glider Pilot Regiment who saw action were killed or wounded. In July 1943, fifty British gliders were dropped by error into the sea off Sicily, and the Allied fleet in Gela harbor accidentally shot down twenty-five planeloads of paratroops passing overhead. Almost all of Merrill's Marauders were "written off."

SHAEF (Supreme Headquarters Allied Expeditionary Force) planners accepted the likely loss of seventy percent of glider troops and half the paratroops in Overlord, the invasion of Normandy. On the eve of D-Day, British Air Marshal Leigh-Mallory tried to stop the airborne operation. While Eisenhower overrode objections, he was in tears on leaving the 101st Airborne Division camp. Fortunately, casualties were under ten percent on D-Day, but in spite of low relative combat time airborne division losses in Europe equaled those of ground divisions. Though never jumping in combat, U.S. Marine Corps parachute units, serving as infantry and as raiders in late summer-autumn 1942, suffered fifty percent losses within a few days of being sent into action. They were the first Marine unit withdrawn in the Solomons campaign, a pattern repeated when the Marine Provisional Parachute Battalion was pulled off "Hellzapoppin' Ridge" Bougainville decimated due to sustained combat with light weapons.[31] As noted earlier, German airborne losses in Holland in 1940 in the first attack against The Hague were forty-two percent of the officers, twenty-eight percent of the enlisted men, and ninety percent of the transport aircraft.

The selection-destruction pattern in respect to other *corps d'élite* has been noted elsewhere; the Rangers at Cisterna and Pointe du Hoe, the Commandos, and the 1st Special Service Brigade were all decimated. The U.S. 10th Mountain Division, the most elite American division in the Second World War in terms of intelligence scores, fitness, and training, suffered five times more casualties preparing for combat than any other American division. Committed in Italy against German mountain troops, the 10th suffered the highest casualties relative to time in combat among U.S. divisions in the campaign. On a larger scale, the U.S. Army Air Forces in Europe and Royal Air Force Bomber Command, both drawn from the upper tiers of men eligible for service, suffered the highest loss rates of any major element in their armed forces, as did the elite submarine and UDT forces in the U.S. Navy, and has already been noted, the *Waffen SS*. In Vietnam, the U.S. Marines' loss rate exceeded that of the Army by a third.

The symbolic power of *corps d'élite* in some nations is used to evoke patriotic fervor or instill awe or fear to foes foreign and domestic. Bureaucratic and political dynamics, however, that shape the organization and doctrine of special forces are kept far from public view over much of the world. As with intelligence and covert operations in general, this is less the case in the United States than other nations, an index of its special moral status among nations to liberally-minded citizens, and of its navet to others. Thus much of the debate surrounding rejuvenation and bureaucratic strengthening of special forces in the Reagan administration has been visible in "open" sources.

As suggested throughout the bibliographic entries, and earlier in this essay, the political flavor of *corps d'élite* and special operations has always been heavy, however much it may have been ignored or minimized by controllers. At the peak of American commitment in the Second Indochina War, in 1968, special operations fell into a trough following revelations regarding the Phoenix program and the "Green Beret murder case."

These episodes seemed to validate fears of many senior U.S. commanders who had resisted creation of special units in the 1950s and expansion in the early 1960s. The tracking down of Che Guevara's band in in 1968 by Colombian troops supported by American special operations elements was seen as a tragedy in some Western circles and in the Communist bloc, and less a triumph in the West than it might have appeared in the mid-1950s or 1980s.

By the middle 1970s, the special elements of the U.S. armed forces had been scaled back. Exposés of the CIA and covert operations in the post-Vietnam era abounded in the popular press and in books and in U.S. congressional hearings on intelligence, although covert operations continued in spite of the Carter administration's softened foreign policy rhetoric.[32] American special counterterrorist units were formed in the wake of an onset of airliner hijackings and terrorist kidnappings in the early 1970s, but other nations' *corps d'élite* scored the major victories in that quasi-war. The failure of the *Eagle Claw* raid in April 1980, aimed at freeing the American hostages in Teheran, was seen by some to dramatize once again the great risks of special operations, and cited by others as symptomatic of a general malaise in the U.S. military system.

At the same time, nonofficial historians, memoirists, and novelists writing on the Second Indochina War began to reveal details of special operations in Vietnam, striking a less sensational balance regarding success and failure. A growing sense of a minimum need for special covert operations capacity in the defense array was also seen on both sides of the American political spectrum, enthusiastic on the right and grudging on the left,[33] as well as within the armed services.[34]

The *Eagle Claw* failure remained a painful memory foremost in a litany of fiascos and mischances, including the *Pueblo*, *Liberty*, and *Mayaguez* incidents, the Son Tay raid of 1972, the Beirut embassy explosion, and floundering among various elites in Grenada in 1983. At the same time, successes of the British SAS in Britain and the Falklands, including reports of operations on the mainland, burnished the tarnished escutcheon of *corps d'élite*, as did French rapid deployment elite forces in Africa, especially Chad, whose upper echelon included many *para* veterans of Algeria. The Libya air raid of 1986 was reported to have been paralleled by SEAL operations ashore. At the same time, the Iran-Contra affair raised the level of public attention regarding covert operations once again, and reports of financial mismanagement in various American *corps d'élite* were alleged.[35]

Phoenix-like, *Eagle Claw* and Grenada had raised consciousness and concern, leading in stages to direct action by the U.S., where Senators Sam Nunn and William Cohen et al. worked with those in the armed services favorably disposed to special operations[36] toward creating something close to a "sixth service"[37] dedicated to special operations and encompassing the Army's Rangers, Special Forces, and DELTA force, the Air Force's air and ground special operations elements, the Marine Corps' Force Recon units, the Navy's SEALs and, apparently, other special elements not readily visible. Special Operations became an Army "career field" in 1983[38] and, in 1987, a full-fledged branch with crossed arrows insignia.

At a higher level, funding was increased and a special operations command (USSOC) was created and headquartered at McDill Air Force Base, Florida (after some controversy), and an Assistant Secretary of Defense for Special Operations and Low-Intensity Conflict slot was created--but not permanently filled.[39] Collaterally, images of the Second Indochina War were rekindled, more in anger than sorrow, including various high-violence films invoking special units' imagery, and paeans to *corps d'élite* in various media.[40] The references in the section on elite forces in the bibliography on Soviet *spetsnaz* reflect growing Western awareness of the Eastern bloc variants of *corps d'élite*. These forces played a professionally skillful and prominent role in the Soviet seizure of the Prague airport in 1968 and as the cutting edge of the Soviet incursion into Afghanistan in 1979.

As this work was being finished in the spring of 1988, *corps d'élite* and special operations had come into view in several critical instances. In the U.S., Iran-Contra indictments and a wave of books engendered by the initial crisis and hearings had brought covert operations in Central America back into public view, and scenes of brutal U.S. Navy SEAL training were being shown on television. In Ulster, SAS' killing of three Irish Republic Army bombers triggered off massive reaction in March and waves of arrests, the British government remained evasive on the questions of counter-terror standing orders. In the Middle East, *Sayaret Matkal* was named in the media as executive agent in the assassination of Palestine Liberation Organization leader in April, 1988. After a long hiatus, skyjacking re-erupted in the holding of a Kuwaiti airliner for over three weeks, without special operations response.

In the Contra affair, in the Persian Gulf where special units have reportedly been deployed, and in Central America, in an age of small wars and quasi-wars, basic tradeoffs implicit in the use of special units and operations remained a relative constant, dangers of misallocation of resources, political consequences of failure, leadership drain and organizational tension balancing against reliability, mystique, and aggressiveness.

When they win, special units and operations may be invaluable, whether they gain publicity or not. But when they visibly fail, the costs are almost always greater politically than their controllers anticipated--strong medicine and strong side-effects. The small print on the bottle should be read very carefully, and easy talk about rationalization emanating from pundits discounted.

OF THE CALLED AND THE CHOSEN: A FINAL DILEMMA

John Erickson has suggested a conceptual triad for focussing analysis of the Soviet military system--technology, doctrine, and style. In the realm of special operations and elite forces, style carries disproportionate weight. This is perhaps more visible in the United States where defense-related discussion is far more a public matter than in other major nations. Other nations may have done a better job of folding this deceptively ephemeral complex of developments into their defense structure and meshing it with policy and strategy.

In any case, both defenses and critiques of low-intensity conflict, in the U.S. at least, share a sense of LIC/Special Ops being a doctrinal stepchild. Thus opposite positions regarding the risk and the efficacy of relying on covert/paramilitary operations have abounded throughout the Cold War, rotating around a polarity of a force structure based on standard firepower heavy units versus more costly specialized, smaller units. However useful such an ongoing exegesis may be in developing alternative perspectives, in sparking creativity, or keeping an adversary in a state of uncertainty, it does not lead to coherence or surety.

Since they began to burgeon in 1940, the political and military harnessing of *corps d'élite* and covert operations has been relatively reactive and peripheral in Western nations. Advocates have had to fight for a forum, and the problem of orchestration is as far from doctrinal balance as was air power in 1938.

Proposals of the "Reform Movement" for lighter, faster forces came to the fore in the late 1970s, arising partly from the sense of frustration in Vietnam and various operations noted earlier, partly from escalating defense costs and technical complexities, and partly from the implications of the 1973 October War, in which electronic warfare and "smart weapons" astounded many participants and observers in their impact.

Another driving force was an awareness in many quarters in the United States and among America's allies that Vietnam had called into serious question the capacity and determination of the United States as the guarantor of various alliances' military line-of-credit. Paralleling that was the rising importance of the Middle East in American geostrategy as the world oil production profile underwent a long-predicted tilt in favor of Third World oil producers.

The focus on "rapid deployment forces" in the wake of the fall of the Shah of Iran in 1978 and subsequent complications led to an uneven consideration of how various special force elements may provide increments to the U.S. order of battle in different scenarios. Again, coherence of thought and doctrine in regard to what each unit might do, could do, or should do in various contingencies remained something of a elusive grail. The tangle of *corps d'élite* in Operation *Urgent Fury* underlined the dysphasia.

In spite of many defense analysts, civilian and military, descrying a poverty of historical awareness in the U.S. officer corps, in the late 1970s through the 1980s the use of history remained uneven and selective. Part of that was due to the fact that civilian policymakers and commanders might not want data to conflict with predisposition, also a problem in the preshaping of intelligence.

Structuring light divisions for quick commitment overseas was a byproduct of U.S. government policy shifting back toward intervention, from the last months of the Carter administration through most of the Reagan administration. Strengthening elite forces themselves and the creation of a stronger bureaucratic overstructure reflected that development in spirit and in fact. Romanticist enthusiasm for dash and maneuver took many forms, e.g., inclusion in U.S. Army doctrine and style in the mid-1970s of main elements of *Wehrmacht blitzkrieg* doctrine and the *Bundeswehr's aufträgstakik*--general mission-type orders.

Much selective history was arrayed by advocates, who tended to overlook many weapon systems based on idealistic visions of mobility and verve that had proven less than panaceas in actual operations. British battle cruisers, designed to "outfight what they couldn't outrun, and outrun what they couldn't outfight," often failed to do either. The Tank Destroyer Corps of World War II, based on lightness and mobility, was perhaps the most short-lived branch in the U.S. Army's history. A series of unpleasant experiences with "jock columns" in the British Eighth Army in North Africa in 1941-42 led to the disbandment of what were essentially combined armed light brigades and a return to standard divisions. Neither their fate nor the unhappy experiences of various *corps d'élite* committed to sustained combat against standard units noted above seemed to modify the optimistic rhetoric of light force proponents.

A special final problem, arising from the mystique of *corps d'élite*, is the quandary that arises from their virtually universal ethos of selection through severe ordeal. Not enough is known about human psychodynamics in many dimensions to give a sense of full confidence about selection templates. In aviation piloting and ground control, in nuclear weapons controller monitoring and other critical areas, the firm use of such criteria has been accepted.

Special force selection is uneven, somewhat more task-related in some cases than others, but rather less scientific than evocative of preliterate tribal banding and bonding, with heavy emphasis on hazing, camaraderie, and forms of athleticism. Training also places a premium on physical stamina, pain-tolerance and pain-seeking, and short-focus, quick reaction mental and motor skills. Traits of character and mind at a premium in covert/special operations, however, may or may not be correlated with those selected through rites-of-passage, as was noted with some sense of perplexity by social scientists in the postmortem of Office of Strategic Services selection techniques, *Assessment of Men*, listed in the bibliography.

Some swagger and bravado of many Western *corps d'élite* may stem from a view of war as a kind of body-contact sport. J.F.C. Fuller tried to point out the implicit anomaly in noting that in Asia, the symbol of war was the bow and arrow, while in Europe, it was the sword.

Basic questions in regard to elite force selection include: Who is recruited and why? What are overt versus hidden or subconscious motives? Are aspirants wholly confident and competent? What are they trying to prove, to others or themselves? Who is turned away or not attracted at all that might be well--or better--suited to certain tasks in the fluid, creative milieux of special operations? Are leaders able to nurture and identify such exotic traits, or does the special hunger for control in military hierarchies create a tension both endemic and irresolvable, except when fate brings an Allenby into conjunction with a Lawrence?

It is easy enough to ask such questions. How they and other questions noted throughout the bibliography and appendices will be answered in discussion and through research rather than learned at the cost of blood, treasure, and political embarrassment remains to be seen.

* * *

Notes

1. For a perspective on the problem, see Robert McQuie, "Military History and Mathematical Analysis," *Military Review*, May 1970, pp. 8-17.

2. For a succinct, colorful account see Cornelius Ryan, *The Longest Day* (New York, 1959), pp. 236-239.

3. See Robert D. Burhans, *The First Special Service Force: A War History of the North Americans*, 1942-44 (Washington, D.C., 1947); Robert H. Adleman and George Walton, *The Devil's Brigade* (New York, 1966); Stanley W. Dziuban, *The United States Army in World War II: Military Relations Between the United States and Canada*, 1939-45 (Washington, D.C., 1959), pp. 265-268.

4. Dziuban, *The United States Army in World War II*, pp. 267-268.

5. *Ibid.*, p. 265.

6. For a precis of the various assaults on the U.S. Marine Corps, see R. D. Heinl, "The Cat with More Than Nine Lives," *U.S. Naval Institute Proceedings*, June, 1954, pp. 659-671.

7. William Slim, *Defeat Into Victory* (New York, 1961), p. 239.

8. Charles F. Romanus, and Riley Sunderland, *Stilwell's Command Problems* (Washington, D.C., 1956), p. 131.

9. U.S. Army Historical Division, *Merrill's Marauders* (Washington, D.C., 1945), pp. 111-112.

10. I. Dontsov, and P. Livotov, "Soviet Airborne Tactics," *Military Review*, October, 1964, pp. 29-33.

11. *Decisive Battles of World War II: The German View*, eds., Hans-Adolf Jacobson, and Jurgen Rohwer, trans. Edward Fitzgerald (London, 1965), p. 131.

12. Gordon A. Harrison, *Cross-Channel Attack* (Washington, D.C., 1951), p. 238; *The War Reports of George C. Marshall, H. H. Arnold and Ernest J. King* (Philadelphia, 1947), p. 170.

13. Bernard L. Montgomery, *The Memoirs of Field Marshal Montgomery* (Cleveland, 1958), p. 266.

14. Martin Blumenson, *Breakout and Pursuit* (Washington, D.C.; Office of the Chief of Military History, 1961), pp. 679-680.

15. Paul-Marie de la Gorce, *The French Army*, trans. Kenneth Douglas (New York, George Braziller, 1963), p. 482.

16. Edgar S. Furniss, Jr., *De Gaulle and the French Army* (New York, George Braziller, 1964), p. 38.

17. *The Goebbels Diaries*, ed. and trans. Louis P. Lochner (New York, 1948), p. 608.

18. *The Politics of the Red Chinese Army: A Translation of Activities of the People's Liberation Army*, ed. Chester J. Cheng (Stanford, 1966), p. 270.

19. *Ibid.*, "Preliminary Conditions Concerning the Establishment of Four Excellence Company Units and Opinions of Several Concrete Problems," p. 392.

20. John Gittings, *The Role of the Chinese Army* (London, 1967), p. 203.

21. Franz Schurmann, *Ideology and Organization in Communist China* (Berkeley, 1966), p. 426.

22. Heinz Höhne, *The Order of the Death's Head* (New York, 1970), p. 28.

23. George H. Stein, *The Waffen SS* (Ithaca, N.Y., 1966), p. 211.

24. Siegfried Westphal, *The German Army in the West* (London, 1951).

25. Jacques Ellul, *The Technological Society*, trans. John Wilkinson (New York: Knopf, 1964); Edward Luttwak, *Coup d'état: A Practical Handbook* (Greenwich, Conn.: Fawcett, 1969), pp. vii-viii.

26. For example, see Albert Parry, *Terrorism: From Robespierre to Arafat* (New York: Vanguard Press, 1976); Yonah Alexander, *International Terrorism* (New York: Praeger, 1976); *Terror and Urban Guerrillas*, ed. Jay Mallin, (Coral Gables, Fla.: University of Miami Press, 1971), p. 148; and Thomas Greene, *Comparative Revolutionary Movements* (Englewood Cliffs, N.J.: Prentice-Hall, 1974).

27. J. Bowyer Bell, "The Profile of a Terrorist: A Cautionary Tale," in *International Terrorism*, ed. Carol Edler Baumann (Milwaukee: Institute of World Affairs, University of Wisconsin-Milwaukee, 1974), p. 12.

28. Bell, "The Profile of a Terrorist: A Cautionary Tale," p. 12.

29. For an extended analysis, see Milton Rokeach, *The Open and Closed Mind* (New York: Basic Books, 1960).

30. Slim, *Defeat Into Victory*, p. 453.

31. John L. Zimmerman, *The Guadalcanal Campaign* (Washington, D.C., 1949), p. 34. See also Henry T. Shaw and Douglas T. Kane, *Isolation of Rabaul: U.S. Marine Corps Operations in World War II*, Vol. II (Washington, D.C., 1961), p. 277.

32. For a review of the reduction and a call for a renaissance, see Noel C. Koch (then Assistant Secretary of Defense), "Why We Must Rebuild Our Special Operations Forces," *Defense*, July, 1983, pp. 8-13.

33. Robert S. Dudney, and Joseph Galloway, "Return of America's Secret Warriors," *U.S. News and World Report*, November 15, 1982, pp. 38-39; and Robert C. Toth, "Elite Troops: Special Forces Trained for Action but Searching for a Mission," *Los Angeles Times Service*, in *Houston Chronicle*, November 25, 1984, sec. 1, p. 28.

34. Michael E. Haas, "Special Forces for Special Problems," *U.S. Naval Institute Proceedings*, (109:7), July, 1983, pp. 110-113.

35. N.a., "Army Investigation Embarrassing to Elite Intelligence Units," *New York Times*, December 1, 1985, p. 16; N.a., "Navy Rescue Team is Subject of Probe," *Washington Post Service*, in *Houston Chronicle*, August 12, 1987, sec. 1, p. 3.

36. Millard Barger, Luanne K. Levens, and Benjamin F. Schemmer, "An Exclusive Interview with General Paul X. Kelley," *Armed Forces Journal International*, (122:1), August, 1984, pp. 61-62, 65-66, 70.

37. Dan Daniel, "U.S. Special Operations: The Case for a Sixth Service," *Armed Forces Journal International*, (123:1), August, 1985, pp. 70, 72, 74-75.

38. Don Hirst, "Special Operations Career Field Approved," *Army Times*, (43:45), June 20, 1983, p. 4.

39. N.a., "New Joint Special Operations, Forces Commends Established," *Army*, (37:6), June, 1987, pp. 89-90.

40. E.g., Robin Moore, "The Green Berets are Back," *Parade*, August 2, 1981, pp. 6-7.

1

Background and Analysis

Alexander, Leo. "War crimes and their motivations: The socio-psychological structure of the *SS* and the criminalization of a society." *Journal of Law and Criminology* (39:3) 1948, pp. 293-296.

Allard, C. Kenneth. "A clear and present danger: Soviet airborne forces in the 1980s." In *Projection of power: Perspectives, perceptions and problems*, eds. Uri Ra'anan, Robert L. Pfaltzgraff and Geoffrey Kemp. Hamden, Conn.: Archon, 1982, pp. 28-48.

Ambrose, Stephen E., and Richard H. Immerman. *Ike's spies: Eisenhower and the intelligence establishment*. Garden City: Doubleday, 1981. 367 pp.

A broad, well-documented survey of Eisenhower's experiences with intelligence as a major commander in World War II, and as president, 1953-61, including substantial data on special operations and *corps d'élite*. 676 refs.

Asprey, Robert. *War in the shadows: The guerrilla in history*. 2 vols. Garden City: Doubleday, 1975. 1,475 pp.

An exhaustive and useful study of irregular warfare, with many sources and cases, this work deals with special units and operations as they relate to guerrilla warfare. A basic point of departure. 796 refs.

Bair, Arthur H., Jr., et al. "Unconventional warfare: A legitimate tool of foreign policy." *Conflict* (4:1) Winter, 1983, pp. 59-82.

A post-*Eagle-Claw*, pre-Grenada perspective by a group of defense professionals which sounds cautions regarding *corps d'élite*; essentially of the "short-leash" school of thought regarding special operations.

Baratto, David J. "Special forces in the 1980s: A strategic re-orientation." *Military Review* (63:3) March, 1983, pp. 2-14.

Barnett, Frank R., B. High Tovar, and Richard H. Shultz, eds. *Special operations in U.S. strategy*. Washington, D.C.: National Defense University, 1984. 326 pp.

A broad-ranging study, close to the crest of the resurgence of covert/special operations during the Reagan administration. The proceedings of a symposium held in 1983 of major figures in the special operations/psy war civilian and military nexus, scholar/analysts for the most part sympathetic, some strongly some rather more cautious. Special operations are most carefully treated by Roger M. Pezelle, "Military Capabilities and Special Operations in the 1980s," pp. 135-152, and by Sam C. Sarkesian, "Organizational Strategy and Low-Intensity Conflicts," pp. 261-289. Also useful is John Dziak, "The Soviet Approach to Special Operations," pp. 95-120. 174 refs.

Beaumont, Roger. "Military elite forces: Surrogate war, terrorism, and the new battlefield." *Parameters: Journal of the U.S. Army War College* (ix:1) March, 1979, pp. 17-29.

An academic analysis of the special operations milieu midway through the Carter administration. 50 refs.

Beckett, Ian F.W., and John Pimlott. *Armed forces and counterinsurgency*. London: Croom Helm, n.d. 232 pp.

Details of British, French, U.S., Uruguayan, Portuguese, Rhodesian, and South African counterinsurgency operations and various special operations and forces. 198 refs.

Bjelajac, Slavko N. "Unconventional warfare in the nuclear era." *Orbis* (4:13) Fall, 1960, pp. 323-337.

Blechman, Barry M., and Stephen S. Kaplan. "Armed forces as political instruments." *Survival* (20:3) July-August, 1977, pp. 169-173.

Burgess, William H. III. "Strategic targeting." *Armed Forces Journal International* (122:8) March, 1985, pp. 66-67, 70, 72, and 74-75.

A critique of American planning and doctrine in respect to development of special operations target data, stressing technical complexity, and noting methods and effects often overridden or ignored by special units and their enthusiasts and controllers; suggests specific new models.

Burnham, James. "The forms of modern warfare." In *The war we are in: The last decade and the next*. New Rochelle: Arlington House, 1967, pp. 240-284.

A veteran conservative essayist whose arguments presaged both the logic and terminology of the renaissance of special and covert operations in the Reagan administration.

Case, Frank B. "Airborne: The tired revolution." *Military Review* (45:8) August, 1965, pp. 86-94.

An analysis of airborne forces' anomalous place in the context of rapidly evolving "air mobile" doctrine and organization and the commitment of *corps d'élite* as the main component in the first phase of the 2nd Indochina War.

Cave Brown, Anthony, ed. *The secret war report of the O.S.S.* New York: Berkeley, 1976. 572 pp.
This work, ranging from the pre-war background of the formation of the Office of Strategic Services to the end of the Second World War, deals mainly with espionage and intelligence. A substantial amount of material on psychological operations and special operations is included, including research and development and training. 39 refs.

Cockerham, William C. "Green Berets and the symbolic meaning of heroism." *Urban Life* (8:1) April, 1979, pp. 94-113.
The U.S. Army Special Forces' ethos analyzed from a social science perspective, using such concepts as extrinsic rewards and the "total institution." 42 refs.

Cohen, Eliot. *Commandos and politicians: Elite military units in modern democracies*. Cambridge: Harvard University Press, 1978. 136 pp.
An essay calling for "deromanticized and depoliticized elites" on the eve of the renaissance of American *corps d'élites*, the emergence of *spetsnaz*, a wave of popular culture romanticization, and international proliferation. 176 refs.

Corson, William R. *The armies of ignorance: The rise of the American intelligence empire*. New York: Dial, 1977. 640 pp.
A survey of the American intelligence community, including special/covert operations and units published in the wake of the dismantling of CIA covert action elements, and the impact of the Church Committee. 130 refs.

Crossland, Roger L. "Unconventional warfare afloat." *U.S. Naval Institute Proceedings* (107:11) November, 1981, pp. 37-41.
A brief for expanded special operations capacity using historical cases.

Deitchman, Seymour. *Limited war and American defense policy*. Cambridge: MIT, 1964. 273 pp.
A brief for strengthened U.S. forces and counterinsurgency on the eve of major U.S. commitment of forces to South Vietnam, dedicated to John F. Kennedy as the champion of the COIN/special unit renaissance. 170 refs.

Dewar, Michael. *Brush fire wars: Minor campaigns of the British army since 1945*. New York: St. Martin's Press, 1984.
A survey of twelve "small wars" with detailing of involvement of *corps d'élite* throughout. 93 refs.

Dyer, Murray. *The weapon on the wall: Rethinking psychological warfare.* Baltimore: John Hopkins Press, 1959. 269 pp.

An essay on the decline of psychological warfare as an element in American foreign policy in the late Eisenhower years.

Eckstein, Harry. "On the etiology of internal wars." *History and Theory* (4:2) Spring, 1965, pp. 133-163.

Engelberg, Stephen. "New book says pentagon failed to inform congress of secret units," *New York Times*, p. 4Y, March 13, 1988.

Evans, Ernest. *Wars without splendor: U.S. military and low-level conflict.* Westport, Conn.: Greenwood, 1987. 160 pp.

Fall, Bernard. "The theory and practice of insurgency and counterinsurgency." In *Last reflections on a war.* New York: Doubleday, 1967, pp. 209-223.

Fall, a French journalist who wrote extensively on the 1st and 2nd Indochina Wars, was sensitive to the political nature of small wars, but his support of the Lansdale approach was falling on deaf ears when he was killed in 1965.

Fallows, James. *National defense.* New York: Vintage Books, 1982. 204 pp.

A critique of U.S. defense policy in the early Reagan administration, couched in the arguments of the Reform Movement, with reference to the logic of forming new light forces and the paradox of high motivation and low retention among elite forces. 208 refs.

Fergusson, Bernard. *The watery maze: A story of combined operations.* New York: Holt, Rinehart and Winston, 1961. 445 pp.

Although weakly documented, Fergusson's study stands as a major work in the field, with much detailing on the role of special operations and units and the oft-neglected dimension of command-and-control. 25 refs.

Fitzsimons, Louise. *Counterinsurgency: The fatal illusion. The Kennedy doctrine.* New York: Random House, 1972, pp. 123-214.

A critique of the basic logic and doctrine of counterinsurgency propounded by Presidents Kennedy and Johnson. 283 refs.

Galula, David. *Counterinsurgency warfare: Theory and practice.* New York: Frederick Praeger, 1964. 143 pp.

The author was a retired French officer and an eyewitness to low-intensity and limited war in several locales. His synthesis of *la guerre revolutionnaire* with emerging American counterinsurgency doctrine provides useful perspectives on special operations and forces doctrine and practice over the next generation in the U.S.

Gavin, James M. *Airborne warfare*. Washington, D.C.: Infantry Journal Press, 1947. 186 pp.

Glover, Michael. *The velvet glove: The decline and fall of moderation in war*. London: Hodder and Stoughton, 1982. 270 pp.

A treatise on the increase in barbaric practices in war in the modern era, including a discussion of Hitler's "Commando Order" of October 18, 1942, and the fate of SAS members captured by the Nazis. 119 refs.

Gravel, Michael, ed. *The Senator Gravel edition--the Pentagon Papers*. 5 vols. Boston: Beacon Press, 1971. 4,100 pp.

Special operations aspects of the Vietnam War are dealt with throughout this working draft of a history of the Department of Defense in the Vietnam era released to the press in 1972 over the opposition of the Nixon administration.

Grivas, George. *General Grivas on guerrilla warfare*. Trans. A.A. Pallis. New York: Praeger, 1962. 109 pp.

A lucid, thoughtful essay on insurgency by a Greek officer who served in both insurgent and counterinsurgent roles, and headed the Greek nationalist EOKA underground on Cyprus, including observation on the role of special units on both sides.

Heymann, H., Jr., and W.W. Whitson. "Specialized forces." In *Can and should the United States preserve a military capability for revolutionary conflict?* Santa Monica: RAND Corporation, 1972, pp. 48-53. [RAND Paper R-940-ARPA]

Written during the period of psychic and physical retrenchment at the end of major U.S. involvement in the 2nd Indochina War, special/covert operations elements, this analysis suggested the major strength of Special Forces lay in their training abilities rather than in direct combat.

Hitler, Adolf. *Mein kampf*. New York: Reynal and Hitchcock, 1941. 993 pp.

A rationale for subsequent Nazi orchestration of *corps d'élite* imagery is laid forth on pp. 750-768.

Hoffman, Bruce. *Commando raids*, 1946-1983. Santa Monica: RAND Corporation, 1985. [RAND Paper N-2316-USDP] 64 pp.

A survey of 100 actions since the end of World War II, with conclusions regarding success (77%); optimum size of action group (15-20); expertise and extended team practice; and the premium value of deception and surprise.

Horan, H.E. "Combined operations." *Journal of the Royal United Services Institution* (43:4) December, 1945, pp. 63-68.

Howarth, Patrick, ed. *Special operations.* London: Routledge and Paul, 1971. 208 pp.

Jenkins, Brian Michael. *Should our arsenal against terrorism include assassination?* Santa Monica: RAND Corporation, 1987. [RAND Paper P-7303] 28 pp.

An essay rejecting the use of assassination in U.S. special operations against terrorists, based on the argument that the United States has a unique moral position to maintain in foreign affairs.

----------. "The U.S. response to terrorism: A policy dilemma." *Armed Forces Journal* (122:9) April, 1985, pp. 39, 41, and 44-45.

A study of terror-counterterror patterns, various possible sources and environments, and suggestions for American policy by a veteran Special Forces officer and RAND analyst.

Jureidini, Paul A., Norman A. la Charit, Bert H. Cooper, and William A. Lybrand. *Casebook on insurgency and revolutionary warfare: 23 summary accounts.* Washington, D.C.: Special Operations Research Office, 1962. 607 pp.

Keynes, Edward. *Undeclared war: Twilight zone of constitutional power.* University Park: Pennsylvania University Press, 1982. 236 pp.

A legal perspective, citing cases and views, on the uncertainties of authority presented by the American constitution's provision for executive-legislative-judicial "separation of powers" in the use of force in foreign policy.

Kitson, Frank. *Low-intensity warfare.* London: Faber and Faber, 1971. 208 pp.

General Kitson's treatise on the use of military and security forces appeared at a peak of political tension in post-Imperial Britain and in northern Ireland, and as ultra-conservative elements moved toward the forming of paramilitary forces, e.g., Column 88. Kitson's views on special operations and special units are set forth on pp. 139, 191-196, and 198. 77 refs.

Knorr, Klaus. "Unconventional warfare: Strategy and tactics in internal political strife." *Annals of the American Academy of Political and Social Science* (341:2) May, 1962, pp. 53-64.

Kupperman, Robert, and William J. Taylor, Jr. "Special supplement: Low-intensity conflict, the strategic challenge."

In *American Defense Annual*, eds. George E. Hudson and Joseph Kruzel. Lexington: D.C. Heath, 1985, pp. 207-222.

A "geostrategic" analysis calling for expansion and rationalization of American special operations/low-intensity warfare capacity during the approach to the Iran-Contra affair. The discussion of "inhibiting factors" foreshadowed the tensions between executive resolve, legisla- tive caution, and the ambiguities of process that emerged in that crisis in late 1986.

Laffin, John. *War annual 1*. London: Brasseys', 1986. 187 pp.

Survey of 29 limited wars in the mid-1980s, including a caustic essay on Grenada.

----------. *War annual 2: A guide to contemporary wars and conflicts*. Oxford: Pergamon Press, 1987. 220 pp.

A survey of wars in progress, mainly insurgencies, including the U.S. air strike against Libya in 1986.

Lewis, Lloyd B. *The tainted war: Culture and identity in Vietnam war narrative*. Westport, Conn.: Greenwood Press, 1985. 193 pp.

An essay, drawing upon American Vietnam war novels and interviews, with numerous allusions to special units and operations. 184 refs.

Linn, Thomas C. The marine corps in special ops. *U.S. Naval Institute Proceedings* (114:2), February, 1988, pp. 48, 50-51.

McMichael, Scott R. *Discussions on training and employing light infantry*. Fort Leavenworth: Combat Studies Institute, c. 1983. [CSI Report No. 8] 21 pp.

An edited transcript of an interview with British officers in 1984, against the background of the U.S. Army's development of Light Divisions for "rapid deployment," and a continuing concern for the questions of motivation and doctrine implicit in raising *corps d'élite*.

Mandel, Robert. *Irrationality in international confrontation*. Westport, Conn.: Greenwood Press, 1987. 176 pp.

Twelve case studies, including the Grenada invasion and the Falklands war.

Marshall, S.L.A. *Men against fire: The problem of battle command in future war*. Washington, D.C.: Infantry Journal Press, 1947. 215 pp.

A controversial landmark study characterizing U.S. infantry battle participation as much lower than popularly perceived in World War II. With its stress on camaraderie, the "buddy" system, and the reluctance of draftees to fight, it raised questions of motivation and reliability to which the raising of all-volunteer forces within the general conscript forces seemed to offer an answer.

Merrill, Bartlett, ed. *Assault from the sea: Essays on the history of amphibious warfare*. Annapolis: U.S. Naval Institute Press, 1983. 453 pp.

A useful overview of amphibious and combined operations, including data on various *corps d'élite*, special operations, and a lengthy bibliography. 327 refs.

Micklous, Edward F. *Terrorism, 1980-1987: A selectively annotated bibliography*. Westport, Conn.: Greenwood Press, 1988. 328 pp. [Bibliographies and indeces in law and political science, No. 8]

Includes a section on "Responses."

Middleton, Drew. "Is gung-ho the way to go?" *New York Times*, April 26, 1981, sec. 4, p. 5.

Miksche, F.O. *Secret forces*. Westport, Conn.: Greenwood Press, 1970. 181 pp.

Miksche's treatise, writen in the shadow of Vietnam, the battle of the Sorbonne, and the spread of wars of national liberation paralleled a wave of theoretical treatises on guerrilla war, on the other side of the barricades.

The military balance. London: International Institute for Strategic Studies, [published annually].

The most cited international order-of-battle; includes data on commando, airborne, and other special units of many nations.

Mockler, Anthony. *The new mercenaries*. New York: Paragon House, 1987. 374 pp.

The first chapter grapples conceptually with origins, definitions, and the overlap with *corps d'élite* of modern mercenary forces. 20 refs.

Moss, Robert. *Urban guerrilla warfare*. London: International Institute for Strategic Studies, 1971. 42 pp.

An analysis in the final phase of direct American involvement in the 2nd Indochina War, including, as an appendix, Carlos Marighella's "Minimanual of the Urban Guerrilla." 48 refs.

Mosse, George. "Fascism and the avant garde." In *Masses and man: Nationalist and fascist perceptions of reality*. New York: Howard Fertig, 1980.

An essay setting the logic of *corps d'élite* within the framework of the Fascist-Nazi worldview.

Motley, James B. *U.S. strategy to counter domestic political terrorism.* Washington, D.C.: National Defense University Press, 1983. [Monograph Series No. 83-2] 136 pp.

----------. "Washington's big tug-of-war over special operations force." *Army* (36:11) November, 1986, pp. 17-24.

Background on political and bureaucratic aspects and a hint of the complex infighting involved in the consolidation of special forces of the various services.

Netanyahu, Benjamin, ed. *Terrorism: How the West can win.* New York: Farrar, Strauss, Giroux, 1986. 234 pp.

The final chapter, an essay by the editor, whose brother was killed commanding the Entebbe raid, has been often quoted and excerpted as perhaps the strongest public policy argument for a broad-gauge use of stringent measures against terrorism, including economic blockade, denial of airline access, and use of military vs. police elements.

Olson, James S., ed. *Dictionary of the Vietnam war.* Westport, Conn.: Greenwood, 1988.

O'Neil, Bard, William R. Heaton, and Donald J. Alberts. *Insurgency in the modern world.* Boulder: Westview Press, 1980. 291 pp.

A survey of counterinsurgencies in Angola, Guatemala, Iraq, Ireland, Oman, Thailand, and Uruguay, including the role of *corps d'élite* and special operations. 74 refs.

Osanka, Franklin Mark, ed. *Modern guerrilla warfare: Fighting communist guerrilla movements, 1941- 1961.* New York: Free Press, 1962. 519 pp.

A compendium of essays on various aspects of insurgency and counter-insurgency that was a major source of concepts as American involvement in Vietnam increased, 1961-1962, and "COIN" became momentarily fashionable. While dated, it is extensive and useful. c. 800 refs.

Paddock, Alfred H., Jr. *U.S. Army special warfare: Its origins: Psychological and unconventional warfare, 1941-1952.* Washington: National Defense University Press, 1982. 221 pp.

A history of U.S. special operations, by a sympathetic but critical veteran of special operations and "psy ops," from U.S. entry into World War II until the middle of the Korean War, apparently to the limits of declassification of basic material at the time of writing, detailing psychological warfare's early link to special operations in U.S. doctrine and organization. 424 refs.

Paget, Julian. *Counterinsurgency operations: Techniques of guerrilla warfare.* New York: Walker & Co., 1967. 189 pp.

Includes brief references to special operations and forces as linked to the Special Branch's Special Operations Volunteer Force in Malaya and "pseudo-gangs" in Kenya. 28 refs.

Paret, Peter. *French revolutionary warfare from Indochina to Algeria: The analysis of a political and military doctrine.* New York: Praeger, 1964. 163 pp.

A succinct account, from an immediate post-war, American academic perspective, of French counterinsurgency doctrine and practice, 1954-62. 138 refs.

Philips, Robert L. *War and justice.* Norman: University of Oklahoma, 1975. 100 pp.

Includes a suggestion that special forces are a preferable modality of force in the context of "just" counterinsurgency, on the basis that they are more likely to be restrained in the application of force.

Powers, B.F. *Is the United States prepared for its most likely conflict?* Santa Monica: RAND Corporation, 1981. [RAND Paper P-6592] 14 pp.

Suggests organization design parameters for intervention forces capable of quick deployment and withdrawal, with high readiness, special logistics, command-and control, special engineering skills, "tunability," high symbolic utility to influence media and foreign opinion, and "mixability" at the last moment before commitment.

Pustay, John S. *Counterinsurgency warfare.* New York: Free Press, 1965. 236 pp.

Includes data on U.S. Army and Air Force special operations elements as the war in Vietnam mounted in earnest. The optimism regarding interdiction and propaganda is poignant in view of subsequent events. 131 p.

Richardson, Robert P., and Stephen Waldron. *An analysis of recent conflicts.* Alexandria: Center for Naval Analyses, 1970. 305 pp.

A study of 380 conflicts 1946-1964 to provide a data base for developing a "process of deterrence and control," and moving toward a theory of "paramilitary warfare."

Rosie, George. *The directory of international terrorism.* New York: Paragon House, 1987. 310 pp.

An useful survey in encyclopedia format of modern, mainly recent terrorist and counterterrorist operations, events, personalities, concepts, and organizations. 100 refs.

Rositske, Harry. "America's secret operations: A perspective." *Foreign Affairs* (53:2) January, 1975, pp. 334-351.

A call for refurbishment of American special/covert operations assets at the post-Vietnam/Watergate nadir.

----------. *The CIA's secret operations: Espionage, counter-espionage and covert action.* New York: Reader's Digest Press, 286 pp.

Rositke's 1975 prospectus for a rationalization and rekindling of U.S. potential in special/covert operations stood athwart the major tides of congressional hostility to such activities in the wake of Vietnam and Watergate. After the Church Committee hearings of 1975 and substantial reduction of CIA covert capacity in the Carter years, a major refurbishment was undertaken from 1981 on.

Sarkesian, Sam C. *The new battlefield: The United States and unconventional conflicts.* Westport, Conn.: Greenwood, 1986. 344 pp.

One of the most important recent works, by a preeminent defense intellectual and Special Forces veteran, laying forth problems and options, with substantial detailing of the structure of U.S. special units and budgeting at the time of major restructuring, on the eve of the Iran-Contra furor.

----------. "Special operations forces in the 1980s." In *The Reagan defense program: An interim assessment*, ed. Stephen Cimbala. Wilmington: Scholarly Resources, Inc., 1986, pp. 93-118.

An overview and conceptual analysis, representative of the school of thought that special operations are essentially an extension of conventional military force structure and mind-set. 33 refs.

Schlachter, David C., and Fred J. Stubbs. "Special operations forces: Not applicable." *Military Review* (58:2) February, 1978, pp. 15-26.

A treatise aimed at defining the role of special operations and forces, viewed marginally in the reorienting of basic U.S. Army doctrine back toward the central European *hauptschlacht* in the aftermath of the 2nd Indochina War. 29 refs.

Scialdone, Antonio. "Naval special operations: A future?" *Military Technology* (8:4) April, 1984, pp. 80-82, 84-85, and 88-89.

An argument that the main element in special units is the selection and rigorous conditioning of people vs. refinement of sophisticated systems.

Shackley, Theodore. *The third option.* New York: McGraw-Hill, 1983. 185 pp.

A retired CIA executive, Shackley laid forth a revitalized U.S. special/covert operations capacity in low-intensity operations, citing selected recent cases.

Shultz, Richard. "The role of external forces in promoting and facilitating international conflict in the third world." In *National security strategy: Choices and limits*, ed. Stephen J. Cimbala. New York: Praeger, 1984, pp. 155-168.

Analysis of Soviet special operations and activities in the Third World, including "active measures," *spetsnaz*, and propaganda.

Smith, Richard Harris. *O.S.S.: The secret history of America's first central intelligence agency*. Berkeley: University of California Press, 1972. 458 pp.

A history of the CIA written in the waning days of the Vietnam war, before Watergate and its major realignment in the mid- and late 1970s.

Snyder, William P., and Roger A. Beaumont. "Military intervention forces." In *Non-nuclear conflicts in the nuclear age*, ed. Sam C. Sarkesian. New York: Praeger, 1980.

An analysis of paradoxes facing the U.S. military in respect to involvement in intervention in the social-political climate of the late 1970s, before the invasion of Afghanistan and the Iranian hostage affair reversed the trends to isolationism engendered by Vietnam and Watergate, and led to a new focussing on elite units in the intervention dynamic. 38 refs.

Tanham, George K., Brian Jenkins, Eleanor S. Wainsten, and Gerard Sullivan. "United States preparation for future low-level conflict." *Conflict* (1:1) Winter, 1978, pp. 1-20.

An academic analysis of special operations potential and organization on the eve of the Iran crisis.

Thayer, Charles W. *Guerrilla*. New York: Signet, 1963. 169 pp.

A brief for strengthening American special forces, by a career Foreign Service officer who served in Yugoslavia in World War II. Short case studies include: the Bay of Pigs; Indochina I; partisan war in Yugoslavia; the Huk rebellion; Malaya and Cyprus. The parallelism of Thayer's "Irregular Warfare Command" with the restructuring of American special operations in the mid-1980s is especially interesting.

Thomas, David. "The implications of commando operations in modern warfare, 1937-1982." *Journal of Contemporary History* (18:4) October, 1985, pp. 689-717.

A selective, sometimes uneven brief for expanded reliance on special units and operations, strongest in its treatment of German *kampftrüppen*, the Brandenburgers et al., and related citations of archival sources. 50 refs.

Thompson, Robert. *Revolutionary war in world strategy*. London: Secker and Warburg, 1970. 171 pp.

A treatise on low-intensity and guerrilla war; Thompson, a principal in the successful counterinsurgency campaign by the British in Malaya, was a consultant to the U.S. government during the Vietnam War and the focus of controversy over how much Malaya resembled--or did not resemble--the conditions in South Vietnam, especially in regard to the "strategic hamlet" program.

----------. "Regular armies and insurgency." In *Regular armies and insurgency*, ed. Thompson. London: Croom Helm, 1979, pp. 9-20.
An analytical essay based on various cases of counterinsurgency in the 20th century, including suggested operating principles.

Thompson, W. Scott, and Andrew B. Walworth. "Domestic political constraints on U.S. military intervention." In *Military Intervention in the third world: Threats, constraints and options*, eds. John H. Maurer and Richard A. Porth. New York: Praeger, 1984, pp. 87-100.

Tillema, Herbert K. *Appeal to force: American military intervention in the era of containment*. New York: Thomas Y. Crowell, 1973. 200 pp.
A survey of U.S. involvement in foreign settings across a spectrum of levels of violence from demonstration to limited war, as American involvement in the Second Indochina War ended. 215 refs.

Trinquier, Roger. *Le temps perdu*. Paris: Albin Michel, 1978. 442 pp.
A memoir by a veteran French colonel and shaper of modern "small war" theory. Especially interesting is the epilogue--"Lesson of Long Experience"--describing the role of mercenaries hired by the insurgent students to fight the *Compagnies Sécurité Républicaine* in the 1968 riots.

U.S. Senate. Select committee of study governmental operations with respect to intelligence operations. Hearings, vol. 7, "Covert action." Washington: U.S. Government Printing Ofice, 1976.

Vlahos, Michael E. "Responding to locally generated contingencies: A gray area in American strategy." In *Military intervention in the third world: threats, constraints and options*, eds. John H. Maurer and Richard A. Porth. New York: Praeger, 1984, pp. 207-230.

Watson, Peter. *War on the mind: The military uses and abuses of psychology*. New York: Basic Books, 1978. 534 pp.

A wide-ranging study of psychological warfare and operations, with extensive data and discussion of special and psychological operations. 694 refs.

West, F.J., Jr. *The strike teams: Tactical performance and strategic potential.* Santa Monica: RAND Corporation, 1969. [RAND Paper P-3987] 18 pp.

An analysis of the experiences of American special units used as a rapid-reaction and raiding force in Vietnam.

Whittier, H. "Soviet special operations/partisan warfare: Implications for today." *Military Review* (59:1) January, 1979, pp. 48-58.

Willems, Emilio. "The emergence of private armies." In *A way of life and death: Three centuries of Prussian-German militarism: an anthropological approach.* Nashville: Vanderbilt University Press, 1986, pp. 116-121.

A recent study based on the burgeoning application of social science methodologies to questions of military phenomenology, organization, and individual and group dynamics.

Wolf, Charles, Jr., and Katherine Watkins Webb, eds. *Developing cooperative forces in the third world.* Lexington: Mass.: Lexington Books, 1987. 224 pp.

A collection of essays on the raising of surrogate forces in various contingencies, including a discussion by Fred E. Haynes and John J. Patrick of airborne troops in strategic quick-reaction roles.

2
Elite Units

Abholt, Dieter. "*Top-polizeisten* in aktion: Der Berliner SEK." *Reader's Digest* [German edition], February, 1983, pp. 43-47.

Adams, Tom. "Special operations forces of the Soviet Union." *Military Intelligence* (8:4) October-December, 1982, pp. 17-18.

Adamson, Ian. *The forgotten men*. New York: Roy Publishers, 1965. 195 pp.
The History of 204 Military (Commando) Mission to China, a three-battalion British Empire force. It was a pseudo-elite unit, like Merrill's Marauders and the 1st Special Service Force, a mix of Australian volunteers and ex-commandos, and marginal personnel. No refs.

Adleman, Robert H., and George Walton (Col.). *The Devil's Brigade*. New York: Bantam, 1966. 239 pp.
A history in popular style of the 1st U.S.-Canadian Special Service Force. No refs.

Airborne operations: A German appraisal. Washington, D.C.: Department of the Army, 1951. [DA Pamphlet No. 20-232] 56 pp.
Written for the European Command on contract by a group of twelve *Wehrmacht* general officers, as part of a major historiographical project with didactic purposes. Includes operational, tactical, and doctrinal material.

Allon, Yigal. *Shield of David*. New York: Random House, 1970. 272 pp.
Includes detailing on the formation of the special night squads and the *Palmach*.

Al-Rayis, Riyad Najib, and Dunia Nahas. *Guerrillas for Palestine*. London: Croom Helm, 1976. 155 pp.

Amphibious warfare headquarters. History of the combined operations organization, 1940-1945. London: Her Majesty's Stationery Office, 1940-1945, 1956. 199 pp.

Andre, John. *World police and paramilitary forces.* New York: Stockton Press, 1985. 245 pp. 53 refs.

Arvidson, John S. (Maj.). "French Foreign Legion update: Where they're at and who they're blowing away." *Soldier of Fortune* (4:2) February, 1978, pp. 59-63.

Avneri, Aryeh. *Sabra commandos.* Tel-Aviv: Olive Books, 1972. 176 pp.

Baker, Lana. *Merrill's Marauders.* New York: Ballantine, 1972. 159 pp.
A brief history of the 5307th Composite Group (Provisional), the only organized ground combat unit of the U.S. Army to fight on the mainland of Asia in World War II.

Barker, A.J. *Waffen S.S. at war.* London: Ian Allen, 1983. 151 pp.
A brief, large-format, illustrated history of the *Waffen SS.*

Beaumont, Roger. *Military elites: Special fighting units in the modern world.* Indianapolis: Bobbs-Merrill, 1974. 251 pp.

Beckwith, Charlie A., and Donald Knox. *Delta Force.* San Diego: Harcourt Brace Javonovich, 1983. 276 pp.
Colonel Beckwith's racy and feisty account of his struggles to create a special force along the lines of the British SAS, with which he served an exchange tour; his service in Vietnam; the creation of Delta Force against the background of mounting concern about terrorism in the mid-1970s; and his command of Operation *Eagle Claw*, the abortive attempt to free American hostages held in Teheran by Shi'ite revolutionary students 1979-81.

Beeson, Colin R. *Glider pilot war.* Manhattan, Ks.: Military Affairs/Aerospace Historian, 1978. 257 pp.
An unedited book of letters from training and the European theater by an American glider pilot of World War II.

Bekker, Cajus. *The K-men: The story of German frogmen and midget submarines.* London: Kimber, 1955. 202 pp.
A short history of the German Special Naval Commandos formed in mid-1944, with technical data on various surface and submersible small-craft, operations, and plans. The work concludes with an account of the plan to drop a thirty-man K-Force group into Berlin to serve as Hilter's bodyguard, as the Nazi air force and SS factions grappled for power in the collapsing Reich.

Belikov, I.G. *Imeni Dzerzhinskogo*. [In the name of *Dzerzhinski*] Moscow: Voenizdat, 1976. 206 pp.

A history of the NKVD *Dzerzhinsky* Division, named for the Polish prime architect of the *Cheka*, primal ancestor of the series of organizations leading to the KGB. Armed elite units of the "organs" have been visible since the Russian Civil War, and currently number about 450,000.

Bellah, James Warner. "The air commando tradition." *Air Force* and *Space Digest* (56:2) February, 1963, pp. 69-79.

Bendman, Sam. "Airborne splashdown: Green Berets hit the water." *Soldier of Fortune* (12:1), January, 1987, pp. 82-85.

Bergen, John D. *Military communications: A test for technology*. Washington, D.C.: Center for Military History, United States Army, 1986. 515 pp.

A volume in the official "U.S. Army in Vietnam series," with extensive detailing on special forces and referencing from official documents.

Berger, Carl, ed. *The United States Air Force in Southeast Asia, 1961-1973: An illustrated account*. Washington, D.C.: United States Air Force, Office of Air Force History, 1984. 383 pp.

Extensive references to special operations, including training and tactics.

Bergôt, Erwan. *La légion*. Paris: Balland, 1972. 339 pp.

A history of the Foreign Legion, in popular format and well-illustrated. Bergot has written extensively on the modern French Army, sub-units of the Foreign Legion, and the airborne, including several titles on the Editions Ballands series "*Corps d'élite*" and the "*Troupes de choc*" series of the Press de la Cité. 40 refs. The following are representative titles:

----------. *Bataillon bigeard: Indochine 1952-1954; Algerie, 1952-1957*. Paris: Presses de la Cité, 1977. 297 pp.

----------. *La légion au combat--Narvik, Bir Hakim, Dién-Bién-Phu. La 13e demi-brigade de la légion étrangère*. Paris: Presses de la Cité, 1975. 317 pp.

----------. *Les paras*. Paris: Balland, 1971. 381 pp.

----------, with John Robert Young. *The French Foreign Legion: The inside story of the world famous fighting force*. London: Thames and Hudson, 1984. 212 pp.

A slick, well-produced, large format illustrated book, similar in essence to Brunon, et al.'s *Le livre d'or. . . .* 32 refs.

Berry, Erick. *Underwater warriors: The story of American frogmen*. New York: McKay, 1967. 152 pp.

Best, Herbert. *The webfoot warriors: The story of U.D.T., the U.S. Navy's underwater demolition teams*. New York: John Day, 1962. 187 pp. 17 refs.

Biggs, Bradley. *The Triple Nickles* (sic). Hamden: Archon Books, 1986. 89 pp.

A brief account of the all-black airborne regiment raised in the U.S. Army in World War II when racial segregation was official policy. The 555th, not committed to combat during the war, faced many of the same obstacles as American black combat air units. 4 refs.

Blair, Clay. *Silent victory: The U.S. submarine war against Japan*. Philadelphia: Lippincott, 1975. 1,072 pp.

Dynamics of an elite force's relations with its parent service are evident throughout, and details on submarine involvement in special operations. 96 refs.

Blassingame, Wyatt. *The U.S. frogmen of World War II*. New York: Random House, 1964. 171 pp.

Blaufarb, Douglas S. *The counterinsurgency era: U.S. doctrine and performance, 1950 to the present*. New York: Free Press, 1977. 356 pp.

A survey of U.S. low-intensity warfare and covert activities from the beginning of the Korean War to the beginning of the Carter administration. Includes details on Vietnam War special operations including 34A operations--secret raids and harassment inside North Vietnam--and hostility in the U.S. Army upper echelon to special operations and forces. 387 refs.

Bloch, Jonathan, and Patrick Fitzgerald. *British intelligence and covert action: Africa, Middle East and Europe since 1945*. Dingle, County Kerry: Brandon, 1983. 284 pp.

A highly critical depiction of the British security-counter-intelligence nexus, including extensive references to the SAS, Special Branch, and related elements.

Böhmler, Rudolf. "Das war die alte fallschirmtrupp" [Those were the old paratroops]. *Deutsches Soldaten Jahrbuch*. Munich: Schild Verlag, 1967, pp. 77-91.

----------, and Werner Haupt. *Fallschirmjäger/paratrooper: Bildband und chronik* [A history in words and pictures]. Revised ed. Dorheim: Verlag Hans-Henning Podzun, 1971. 280 pp.

An English-German illustrated survey of the history of German airborne forces, including a brief section on the *Bundeswehr*'s 1st *Luftlande* (Air-landing) and 1st *Fallschirmjäger* (Paratroop) Divisions.

Bolt, David. *Gurkhas*. New York: Delacorte Press, 1967. 128 pp.
A brief popular-format history of the Gurkhas. 12 refs.

Bonnecarriere, Paul. *Qui ose vaincra: Les parachutistes de la France libre* [Who dares, wins: The parachutists of free France]. Paris: Fayard, 1971. 468 pp.

Borghese, J. Valerio. *Sea devils*. Trans. James Cleugh. London: Andrew Melrose, 1952. 261 pp.
An account of the creation of the Italian underwater warfare units by one of its founders.

Bowman, John S. *The Vietnam war: An almanac*. New York: World Alamanac Publications, 1985. 512 pp. 65 refs.

Boyd, Robert S. "*Spetsnaz*: Soviet innovations in special forces." *Air University Review* (38:1) November-December, 1986, pp. 63-69. 17 refs.

Bragadin, Marc' Antonio. "The deeds of the naval assault teams." In *The Italian navy in World War II*, trans. Gale Hoffman. Annapolis: U.S. Naval Institute Proceedings, 1957, pp. 274-296.
Substantial material on the 10th *Motoscafi Antisubmergilio*, the unit that opened up many avenues in underwater warfare in World War II, including underwater demolition. Much of their equipment and experience was passed on to the British after Italy's surrender in 1943.

Breuer, William B. *Drop zone Sicily: Allied airborne strikes, July 1943*. Novato: Presidio Press, 1983. 212 pp.
A popular account, focussed primarily on American operations in Operation HUSKY, drawn heavily from interviews with 82nd airborne division veteran officers.

Brockdorff, Werner. *Geheimskommandos der zweiten weltkrieges*. München-Weis: Verlag Welsermühl, 1967. 446pp.
A somewhat impressionistic and selective survey of World War II special forces, subtitled "History and Operations of the Brandenburgers, of the English Commandos and SAS units, of the American Rangers and Soviet Secret Services." Somewhat confusing in its use of U.S. unit identities, its main focus is on the Brandenburgers and other German special (not necessarily elite) units. 24 refs.

Brooks, William. "An American 82nd airborne vet provides an inside look: The French Foreign Legion today--part I." *Soldier of Fortune* (3:4) July, 1978, pp. 31-35.

----------. "The French Foreign Legion today--part II." *Soldier of Fortune* (3:5) September, 1978, pp. 36-42.

Brown, Asley, ed. *The Green Beret: U.S. special forces from Vietnam to Delta Force*. New York: Villard Books, 1986. 96 pp.

A concise, well-detailed and illustrated history, ranking very high in usefulness among the "pop" genre of military elite/special operations literature. 5 refs.

Brown, F.C. "SEAL team operations--Vietnam, 1967-1971." *Military Journal*, September-October, 1977, pp. 42-43.

Brown, Robert K. (Lt. Col.). "Rhodesia is ready. Its elite forces are on red alert." *Soldier of Fortune* (1:3) Summer, 1976, pp. 30-33.

Brunon, Jean, Georges-R. Manue, and Pierre Carles. *Le livre d'or de la légion étrangère (1831-1976)*. Paris: Charles-Lavauzelle, 1976. 508 pp.

Essentially a reproduction with additions of the 1958 edition of the "Gold Book of the Foreign Legion." (The most elegant and "collectible" is the 1931 centennial edition.) A broad-gauged, and profusely illustrated work, with special sections on commanders-in-chief and essays on social psychology, road-building, civil works, folklore, the 1931 centennial, songs, etc. 251 refs.

Buchheim, Hans. "The S.S.: Instrument of domination." In *Anatomy of the S.S. State*, eds. Helmut Krausnick et al., trans. Richard Barry. New York: Walker, 1967, pp. 127-397.

----------. "Die SS-trüppen." In *S.S. und polizei im NS-stäat*. Duisdorf: Selbstverlag der Studien-gesellschaft for Zeitproblems, 1982, pp. 165-186.

A portion on the combat units of the *SS-Schützstaffel*, in a detailed study of the evolution of *SS* focussing mainly on the security and police elements. 14 refs.

Burhans, Robert (Lt. Col.). *The first special service force: A war history of the North Americans, 1942-1944*. Washington: Infantry Journal Press, 1947. 376 pp.

A history of the Canadian-American 1st Special Service Force raised to raid Norwegian industry, and diverted to the Aleutians and Italy; includes a final chapter on the 474th Infantry Regiment (Separate) formed from the remnants of the 1st SSF, the Rangers, some airborne elements, and fillers in late 1944. No refs.

Buss, Philip H., and Andrew Mollo, eds. *Hitler's Germanic legions: An illustrated history of the Western European legions with the S.S., 1941-1943*. London: MacDonald and Jane's, 1978. 144 pp.

Illustrated works are useful in developing a sensitivity to the imagery of *corps d'élite*, a major element in the development and dissemination of their mystique, in recruiting and in generating favorable and hostile reaction among their own and opposing forces, and among various civilian groups.

Callinan, Bernard. *Independent company: The 2/2 2/4 Australian independent companies in Portuguese Timor, 1941-1943*. London: Heinemann, 1953. 235 pp.

Calvert, Michael. *Chindits: Long range penetration*. New York: Ballantine, 1973. 159 pp.

A short history of British long-range penetration operations in Burma, 1943-44, by one of Wingate's lieutenants.

Carlson, Evans F. *Twin stars of China*. New York: Dodd, Mead, 1940. 331 pp.

Carlson, a retired U.S. Marine officer, visited China as the Japanese launched their major attacks in 1937, and observed both Nationalist and Communist elements. He drew heavily on organization and training methods of the latter's Eighth Route Army described in these accounts when he later led the U.S.M.C.'s 2nd Raider Battalion in World War II.

Cerff, Karl. *Waffen SS im wehrmachtbericht* [The Waffen SS in Wehrmacht records]. Ösnabruck: Munin-Verlag, 1971. 115 pp.

A collection of extracts from *Wehrmacht* (Nazi armed forces) war diaries and reports dealing with the *Waffen SS*. 10 refs.

Cervia, Mario. *The hollow legions: Mussolini's blunder in Greece, 1940-1941*. Trans. Eric Mosbacher. New York: Doubleday, 1971. 373 pp.

Extensive detailing on Italian forces in the Albanian and Greek campaigns of 1939-41, including *Alpini, Bersaglieri*, and Blackshirt militia.

Chakrabandu, M.C. Kardwik. "Force 136 and the Siamese resistance movement." *Asiatic Review* (163:2) April, 1947, pp. 168-170.

Chaplin, Gordon. "Toughing it out: On the beach, in the bars with the corps d'elite called the SEALs." *Washington Post Magazine*, February 19, 1978, pp. 9-11, and 14-17.

Cheng, J. Chester, ed. *The politics of the Chinese Red Army: A translation of the bulletin of activities of the People's Liberation Army*. Stanford: Hoover Institution Publications, 1966, pp. 270-271, and 387-392.

Details on the Four Excellence companies, model units, prizes, and honor rolls.

Childers, James Saxon. *War eagle: The story of the Eagle Squadron*. New York: D. Appleton-Century, 1943. 350 pp.

A racy wartime history of the American volunteers serving with the British Royal Air Force in the early years of World War II.

Churchill, Winston S. *Their finest hour*. Boston: Houghton-Mifflin, 1949. 751 pp.

Churchill's role in the formation of the Commandos and combined operations is presented herein from his perspective. The resistance of regular force commanders is documented as well.

Cockerham, William C. "The Green Beret reservist and his attitudes toward civil disorder." *Pacific Sociological Review* (18:3) July, 1985, pp. 279-291.

Collins, J.M. *Green Berets, SEALs and spetsnaz*. London: Pergamon, 1987. 186 pp.

A comparative study of American and Soviet special operations forces aimed at improving U.S. capacity, by the Congressional Research Service at the behest of the House Armed Services Committee. 322 refs.

Conner, Albert Z., and Robert G. Poirier. *Red Army order of battle in the great patriotic war*. Novato: Presidio Press, 1985. 408 pp.

Includes sections on honorific titles for corps and divisions and guards divisions.

Cook, Graeme. *Commandos in action*. New York: Taplinger, 1973. 175 pp.

A popular account of the origins of the British Commandos and selected operations in North Africa, St. Nazaire, Dieppe, and a final chapter on Skorzeny. No refs.

Cookridge, E.H. *Set Europe ablaze*. New York: Thomas Y. Crowell, 1967. 410 pp.

A well-referenced popular history of Special Operations Executive which deals at length with selection and training, special units, political warfare, links between political warfare, propaganda, espionage, sabotage, paramilitary operations, communications, and logistics. 215 refs.

Cooper, Jeff. "GSG 9: The philosophy." *Soldier of Fortune* (5:3) March, 1980, pp. 43-45.

Corta, Henry. *Les bérets rouges*. Paris: Amicale des Anciens Parachutistes SAS, 1952. 3rd ed. 329 pp.

Cranshaw, Nancy. *The Cyprus revolt: An account of the struggle for union with Greece*. London: George Allen and Unwin, 1978. 447 pp.

Includes material on counterinsurgency/special operations of Royal Marine Commandos and airborne forces. 613 refs.

Craven, Wesley Frank, and James Lea Cate. *The army air forces in World War II*. Vols. 1-6. Chicago: University of Chicago, 1948-1953.
Material on combat air crew selection, training, and operations in various elite forces and special operations, including airborne airlift, the AVG, the 509th Composite Unit, and Air Commandos. Sourced mainly from U.S. Army Air Force official records.

Crookenden, Napier. *Airborne at war*. New York: Scribners, 1978. 144 pp.
A large-format illustrated volume, focussing on specific airborne operations, German *fallschirmjäger* at Eben Emael in 1940, the U.S. drop on Corregidor, and the Anglo-American paradrop phase of crossing the Rhine in 1945.

Crouse, G.B. "Spike teams, hatchet forces, SLAM companies." *Soldier of Fortune* (12:12) December, 1987, p. 66.
A survey of U.S. Special Operations Group 35 (Ground Studies Group) and its constituent elements in the late 1960s in the 2nd Indochina War.

Dank, Milton. *The glider gang*. New York: J.B. Lippincott, 1977. 273 pp.
A brief history of American glider operations in World War II based mainly on literally dozens of interviews with veterans. 68 refs.

Darragh, Shaun M. "Rangers and special forces: Two edges of the same dagger." *Army* (127:12) December, 1977, pp. 14-19.
A brief for expanded American special forces/operations, including an overview of various elements in the 2nd Indochina War.

Davies, Howard P. *British parachute forces, 1940-1945*. New York: Arco, 1974. 32 pp.
A brief illustrated work.

Dean, David J. *The air force role in low-intensity conflict*. Maxwell Air Force Base: Air University, 1986. 127 pp.
Includes an overview of air power in small wars; a detailed case study (Morocco); an overview of U.S. Air Force Special Operations evolution; and prognostications. 155 refs.

Deighton, Len. "The private armies." In *The Sunday Times book of Alamein and the desert war*, ed. Derek Jewel. London: Sphere Books, 1967, pp. 130-137.

A survey of the activities of the Long Range Desert Group, SAS, and Popski's Private Army, including tactics and equipage.

Devlin, Gerard M. *Paratrooper! The saga of U.S. Army and Marine parachute and glider combat troops during World War II.* New York: St. Martin's Press, 1979. 717 pp.

A well-referenced, adulatory history. 191 refs.

----------. *Silent wings: The saga of the U.S. Army and Marine Corps combat glider pilots during World War II.* New York: St. Martin's, 1985. 410 pp.

A detailed, well-sourced, and illustrated account of a mode of airborne warfare that was on balance more successful in several principal operations than the use of parachutists, but which lacked the mystique and glamour of the latter. 111 refs.

Dewar, Michael. *The British army in northern Ireland.* London: Arms and Armour Press, 1985. 270 pp.

Includes data on various *corps d'élite* on security tours in Ulster. 53 refs.

Deysdale, D.B. "Special forces." *Marine Corps Gazette* (32:6) June, 1954, pp. 47-53.

Divilkovksy, S., and I. Ognetov. "The face of special warfare." In *The road to victory: The struggle for national independence, unity, peace, and socialism in Vietnam.* Moscow: Progress Publishers, 1978, pp. 97-102.

A Soviet perspective on U.S. special operations in the 2nd Indochina War.

Dobson, Christopher, and Ronald Payne. *Counterattack: The West's battle against the terrorists.* New York: Facts On File, 1982. 198 pp.

Detailed descriptions of antiterrorist organizations of the U.S., U.K., F.R.G., Israel, Holland, France, Italy, Spain, and private commercial activities. 42 refs.

Dontsov, I., and P. Livotov. "Soviet airborne tactics." *Military Review* (54:10) October, 1964, pp. 29-33.

Doyle, Michael R. "Airborne vigil: China's paras ready on command in Taiwan." *Soldier of Fortune* (11:10) October, 1986, pp. 46-51.

Drenkowski, Dana. "America's underwater elite: U.S. Navy SEALs--part one." *Soldier of Fortune* (4:3) March, 1979, pp. 38-43.

----------. "America's underwater elite: U.S. Navy SEALs--part two--conclusion." *Soldier of Fortune* (4:4) April, 1979, pp. 48-53.

Duncan, Donald. *The new legions*. New York: Random House, 1967. 275 pp.

Duncan, later editor of *Ramparts* magazine, a major forum of anti-war radicalism during the Vietnam war, had served in Special Forces in the early phase of U.S. involvement. His critique appeared on the eve of a series of scandals that wracked the U.S. military establishment as U.S. public opinion shifted away from support of U.S. involvement.

Duncan, N.W. *79th armoured division: Hobo's funnies*. Windsor: Profile, 1972. 224 pp.

A popular history of the unit of specialized armored vehicles that kept casualties on British and Canadian landing zones much lower than on the American beaches. An example of the degree to which specialized technology and exotic skill has been a key element in the definition of elite and "special," in spite of the persistence of the heroic hand-to-hand warrior perceived by many outside the realm of special operations.

Dwyer, John B. "SEAL saga: The history of our webfooted warriors." *Soldier of Fortune* (10:10) October, 1985, pp. 52-55.

Dyer, Murray. *The weapon on the wall: Rethinking psychological warfare*. Baltimore: Johns Hopkins Press, 1959. 269 pp.

Edwards, Roger. *German airborne troops, 1936-1945*. Garden City: Doubleday, 1974. 160 pp.

Elliott, Charles Brandon. "Soviet black berets: The elite force of the Soviet naval infantry." *Soldier of Fortune* (5:2) February, 1980, pp. 60-68.

Ellis, Gerald. "*Groupement des fusiliers marins commandos: French naval commandos*." *Soldier of Fortune* (11:8) August, 1986, pp. 68-73.

English, Adrian J. *Armed forces of Latin America: Their histories, development, present strength and military potential*. London: Jane's Publishing Co., 1984. 490 pp.

Includes extensive data on various elite/special units. 174 refs.

Erickson, John, Lynn Hansen, and William Schneider. *Soviet ground forces: An operational assessment*. Boulder: Westview, 1986. 267 pp.

Airborne and *spetsnaz* units are discussed in the broader context of Soviet combined arms doctrine. 179 refs.

Eshel, David. *Elite fighting units*. New York: Arco Publishing, 1984. 209 pp.

A well-illustrated, selective survey of elite forces and special operations since World War II, with emphasis on Britain, the U.S., and Israel. Some sections read rather like recruiting brochures. Factual detail uneven, e.g., Popski, the 10th Mountain Division, the 509th Composite Group, Delta Force, and the Alamo Scouts are not mentioned.

Faldella, Emilio. *Storia delle truppe alpini, 1872-1972* [History of the Alpine troops, 1872-1962). 3 vols. Milan: Cavalloti Edizioni Landoni, 1972. 2,224 pp.
A highly detailed survey of campaigns, weapons, colors, etc. of the Italian mountain troops.

Falwell, Byron. *The Gurkhas*. New York: W.W. Norton, 1984. 317 pp.
The best recent survey history of the Nepalese contract soldiers who have served both British and Indian governments. 68 refs.

Fane, Francis Douglas, and Don Moore. *The naked warriors*. New York: Appleton-Century Crofts, 1956. 308 pp.
A survey history of American underwater demolition teams' world war tactics, operations, and training, at a time when the U.S., as in many other technical areas, was well behind the state of the art. Poorly referenced, but readable.

Feury, George. *Le 1er regiment de chasseurs parachutistes* [The 1st Regiment of Parachute Light Infantry]. 3 vols. Paris: Charles-Lavauzelle, 1982, 1984, 1985. Vol. 1, 128 pp.; vol. 2, 112 pp.; vol 3., 135 pp.
The heavily illustrated, large format volumes are valuable for gaining a visual sense of the *para* mystique, and the convoluted trail that led to the crises and tragedies of the late 1950s and early 1960s. Mid-1930s links with the Red Army are traced out.

Flamand, Roger. *Paras de la France libre* [Paras of Free France]. Paris: Presses de la Cité, 1976. 317 pp.
A survey of the Free French airborne forces in the Second World War. No refs.

Foot, Michael. *S.O.E. in France*. London: Her Majesty's Stationery Office, 1966. 550 pp.
The definitive, detailed history of Special Operations Executive, the structure that oversaw most of Britain's espionage and special operations in World War II. In many operations, it was closely linked with the American O.S.S. Written from official documents, it has substantial detailing on training and tactics, and the role of the SAS, Jedburghs, and on special operations, psychological warfare, etc. The author's essay on sources indicates what he was not allowed to use, for

the use of future historians, as well as serving as a faint *caveat* and *apologia*. 150 refs.

Fricaud-Chagnaud, Georges (Gen.). *Origins, capabilities and significance of the Force d'Action Rapide*. Paper prepared for "Prospects for Effective Conventional Defense in Europe," International Security Studies Core Seminar, the Wilson Center, Washington, D.C., October 30, 1984. 33 pp.

A description of French rapid deployment forces in the mid-1980s. 12 refs.

Futrell, Robert F., and Martin Blumenson. *The U.S. Air Force in Southeast Asia: The advisory years to 1965*. Washington: Office of Air Force History, 1981. 398 pp.

Extensive references on various special units and operations, including Air Commandos, Jungle Jim (4400th Combat Crew Training Squadron), American and South Vietnamese rangers and special forces. 195 refs.

Gal, Reuven. *A portrait of the Israeli soldier*. Westport, Conn.: Greenwood, 1986. 273 pp.

A wide-ranging survey including substantial data on motivation, selection and training, and elite forces in the Israeli Defense Forces, including detailed information on fighter pilots, the airborne, naval special forces, the *Palmach*, reconnaissance forces, submariners, the Golani Brigade and the Armored Corps, as well as the system of decorations and a short history of the evolution of the IDF, and the inner core of regulars, the *Sherut-Keva*. 179 refs.

Galey, John H. "Bridegrooms of death: A profile study of the Spanish Foreign Legion." *Journal of Contemporary History* (4:2) April, 1969, pp. 48-63.

A sociological-historical view of the *Legion de extranjeros* from its founding in the early 1920s to the 1960s.

Gann, L.H., and T.H. Henrikson. *The struggle for Zimbabwe: Battle in the bush*. New York: Praeger, 1984. 154 pp.

Substantial material on Rhodesia as a "counterinsurgency state," including discussions of elite-standard force "bifurcation" and organization and tactics of Rhodesian *corps d'élite*. 58 refs.

Garrett, Richard. *The raiders: The elite strike forces that altered history*. New York: Van Nostrand Reinhold, c. 1980. 220 pp.

Popular, illustrated-format account of selected raids from the American Civil War to Entebbe. 45 refs.

Gaudet, Ken. "SADF's pathfinders: Yank para pulls through." *Soldier of Fortune* (11:5) May, 1986, pp. 50-53.

Generous, Kevin, M. *Vietnam: The secret war.* New York: Gallery Books, 1986. 192 pp.

Extensive, unreferenced details on French, North and South Vietnamese, American, and Allied special operations and units 1954-1975. (Illustrations and captions are not on a par with the text.)

Geraghty, Tony. *This is the S.A.S.: A pictorial history of the Special Air Service Regiment.* New York: Arco Publishing, 1983. 156 pp.

A survey history of the Special Air Service Regiment from 1941 to the late 1970s. The graphic design and printing is not of the quality to offset the lack of actual action photographs. Much interesting detail, nonetheless. 10 refs.

----------. *Who dares wins: The story of the Special Air Service.* London: Arms and Armour, 1980. 249 pp.

The definitive history to date; World War II origins are presented briefly in the introduction, with emphasis on the SAS since 1950 in Malaya, Borneo, Aden, the Yemen, Oman, Ireland, and Mogadishu; includes such topics as selection, training, doctrine and operation, and variants of the SAS in countries other than the U.K. 65 refs.

Gheline, Patrick de. *Les corps-francs 39-40*. Paris: Presses de la Cité, 1983. 271 pp.

A survey of the operations of volunteer units raised within various French army units early in World War II to perform scouting and raiding functions. The author suggests they were main actors in the fighting of 1940, suffering heavily in protecting the withdrawal of the French Army. Appraisal has been hampered by rigid French security rules on documents, and the loss and destruction of *Corps-francs* records during the 1940 debacle. 49 refs.

Gillette, J. G. "What's French, foreign and fights all over?" *The London Times Magazine*, October 7, 1985, pp. 14-16.

A Sunday supplement on the French Foreign Legion.

Gittings, John. *The role of the Chinese army.* London: Oxford University, 1967. 331 pp.

Includes data on selection and designation of elite forces in the People's Land Army in the post-civil war period. 161 refs.

Glantz, David M. (Lt. Col.). *The Soviet airborne experience.* Fort Leavenworth: Combat Studies Institute, 1984. [Combat Studies Institute Research Survey No. 4.] 211 pp.

A thorough review of the development of Soviet airborne forces in the early 1930s, the political-organizational turbulence of the late 1930s, wartime operations, and a brief discussion of postwar developments. 277 refs.

Glenn, John (Sfc.). "Special forces desert training." *Soldier of Fortune* (1:2) Spring, 1976, pp. 32-39.

Gordon, John. *The other desert war. British special forces in North Africa, 1940-43.* Westport, Conn.: Greenwood, 1987. 241 pp.

Govan, Thomas P. *History of the 10th light division (alpine).* Washington, D.C.: Historical Section, Army Ground Forces, 1946. [Study No. 28.] 14 pp.

One of a series of studies on the organizational evolution of the 10th Mountain Division in World War II, e.g., Study No. 23, Training for Mountain and Winter Warfare. Includes mainly references to basic documents. 42 refs.

Graves, Jim. "SOG's secret war: SOF opens top-secret Pentagon file." *Soldier of Fortune* (6:6) June, 1981, pp. 39-45.

"Greenham defenses copied for spetsnaz training." *Jane's Defense Weekly* (5:3) January 25, 1987, pp. 84-85.

Includes geographic location and organizational description of Soviet GRU (military intelligence) *spetsnaz*.

Gregory, Barry. *British airborne troops, 1940-45.* New York: Doubleday, 1975. 160 pp.

Hampshire, A. Cecil. *The secret navies.* London: William Kimber, 1978. 272 pp.

A history of three of many British naval special forces in World War II: the 15th MGB (motor gunboat) Flotilla, which landed and picked up agents in Europe; the Royal Marine Boom Patrol Detachment (the "Cockleshell Heroes"; and 30 AU--Intelligence Assault Unit--including the Forward Interrogation Unit, which served mainly on land in northwest Europe. 62 refs, including Public Record Office file citations.

Hansen, James. "Soviet vanguard forces--*spetsnaz*." *National Defense* (70:416) March, 1986, pp. 28-32, and 36-37.

Hardesty, Von. *Red phoenix: The rise of Soviet air power.* Washington: Smithsonian Institution Press, 1982. 288 pp.

An exhaustive study of the Soviet air forces in World War II, including training, aces, and other relevant details. 900 refs.

Harrington, Joseph D. *Yankee samurai: The secret role of Nisei in America's Pacific victory.* Detroit: Pettigrew Enterprises, 1979. 383 pp.

Includes material on the raising of volunteers for the 100th Battalion/442nd Regimental Combat Team and military intelligence field interpreter-translation duty from among the youth in the Japanese-American concentration camps in World War II. Enthusiastic in tone.

Harrison, Derrick I. *These men are dangerous: The Special Air Service at war.* 2nd ed. London: Cassell, 1957. 240 pp.

Hastings, Max. *Das Reich: The march of the 2nd S.S. panzer division through France.* New York: Holt, Rinehart and Winston, 1981. 264 pp.

Drawn heavily from interviews, this account includes *Das Reich*'s contacts with various Allied elite forces including elements of SOE and the SAS in the wake of the Normandy invasion in 1944.

Haugland, Vern. *The eagle's war: The saga of the Eagle Squadron pilots, 1940-1945.* New York: James Aronson, 1982. 234 pp.

Hausser, Paul. *Soldaten wie andere auch: Der weg der Waffen SS* [Also soldiers like the others: The way of the Waffen SS]. Osnabruck: Munin Verlag, 1966. 371 pp.

Includes a 140-page source document collection.

----------. *Waffen-SS im einsatz* [Waffen SS in Operations]. Gottingen: Plesse Verlag K. W. Schutz, 1953. 272 pp.

Hawkins, Philip. *Blackburn's headhunters.* New York: Norton, 1955. 326 pp.

An account of the formation of a shadow army, the U.S. Armed Forces in the Philippines (USAFIP), during the Second World War after Japanese occupation, focussing on the activities of Donald Dunwoody Blackburn who, as a general officer, played a major role in U.S. special operations in the 2nd Indochina War. No refs.

Heifenman, Ron. *Flying Tigers: Chennault in China.* New York: Ballantine, 1971. 160 pp.

A brief history of the American Volunteer Group, 1941-42. 10 refs.

Heilbrunn, Otto. *Warfare in the enemy's rear.* New York: Praeger, 1963. 231 pp.

A survey history of special operations in World War II and in the Cold War, with substantial detailing on the Brandenburgers.

Heilman, Ludwig. "Fallschirmjäger auf Sizilien" [Paratroops on Sicily]. *Deutscher Soldatenkalendar.* Munich: Schild Verlag, 1959, pp. 56-64.

An account of German airborne operations in the Sicilian campaign of 1943.

Heinl, Robert Debs, Jr. *Soldiers of the sea: The United States Marine Corps, 1775-1962*. Annapolis: U.S. Naval Institute, 1962. 692 pp.

Contains substantial detailing on various elite sub-elements of the U.S. Marine Corps. 656 refs.

Hickey, Gerald Cannon. *Sons of the mountain: Ethno-history of the Vietnamese central highland to 1954*. New Haven: Yale University Press, 1982. 488 pp.

Includes details on French-raised paramilitary forces.

Hickey, Michael. *Out of the sky: A history of airborne warfare*. London: Mills and Brown, 1979. 286 pp. 31 refs.

Hills, R.J.T. *Phantom was there*. London: E. Arnold, 1951. 159 pp.

A history of the G.H.Q. Liaison Regiment, a special signals unit (later emulated by the U.S. Army) which transmitted battle information to higher echelons, bypassing the chain of command.

Hilton, Frank. *The paras*. London: British Broadcasting Corporation, 1983. 248 pp.

A volume designed to complement a BBC television series, with background on airborne development and the British army parachute regiment, focussing on a group in training. 25 refs.

Höhne, Heinz. *The order of the death's head: The story of Hitler's S.S.* New York: Coward-McCann, 1970. 690 pp.

One of the major studies of "a bizarre nonsensical affair . . . the most fantastic association of men imaginable," tracing roots to Himmler's and Hitler's medieval romanticism, World War I storm troops, the Jesuits, and as a counter to the SA Brown Shirts. 3,661 refs.

Hough, Frank O., Verle E. Ludwig, and Henry I. Shaw. *History of the U.S. Marine Corps operations in World War II*. 3 vols. Washington, D.C.: Historical Division, United States Marine Corps; vol. 1, 1958, 439 pp.; vol. 2, 1963, 632 pp.; vol. 3, 1968, 945 pp.

These volumes contain substantial references and sources on Marine Corps raider and parachute units.

Hoyt, Edwin P. *Airborne: The history of American parachute forces*. New York: Stein and Day, 1979. 228 pp. 12 refs.

----------. *The kamikazes*. New York: Arbor House, 1983. 333 pp. A popular history of the special attack corps. 33 refs.

Hubbell, John G. "Supercommandos of the wetlands." *Reader's Digest* (90:542) June, 1967, pp. 49-54.
A popular view of SEALs in Vietnam.

Hunter, Charles N. *Galahad.* San Antonio: Naylor, 1963. 233 pp.
Galahad Force was the designation for U.S. ground forces in north Burma in 1944, more popularly known as "Merrill's Marauders."

Hure, R. et al., eds. *L'armée d'Afrique, 1830-1962.* Limoges: Charles-Lavauzelle, 1977. 485 pp.
A large-format illustrated overview of French colonial forces whose eliteness varied, including the Foreign Legion, *Chasseurs d'Afrique, Spahis, Tirailleurs Morocains,* and *Goums.*

Husemann, Friedrich. *Die gute glaubens waren: Geschicht der SS-polizei-division* [Those were good memories: History of the *SS* police division]. Osnabrück: Munin Verlag; vol. 1 (1939-1942), 1971, 472 pp.; vol. 2 (1943-1945), 1972, 703 pp. 42 refs.

Hussain, M. *The Palestine Liberation Organization: A study in ideology, strategy and tactics.* Delhi: University Publishers, 1975. 156 pp.
A survey of the development of anti-Israeli *Fedayeen* guerrilla/commando forces since 1948.

Huston, James. *Out of the blue: U.S. army airborne operations in World War II.* West Lafayette: Purdue University Press, 1972. 327 pp.

Inoguchi, Rikehei, Tadashi Nakajima, and Roger Pineau. *The divine wind: Japan's kamikaze force in World War II.* Annapolis: U.S. Naval Institute, 1958. 240 pp.
The English-language account closest to the Japanese idiom; a grisly fore glimpse of the effect of "smart" weapons against massed fleets. Substantial statistical analysis and organizational detailing of perhaps the penultimate modern elite force.

Isby, David C. "Ivan's elite. How will the Soviets deploy their special forces in 1981?" *Soldier of Fortune* (5:10) October, 1980, pp. 66-81.

----------. "The spetsnaz in Afghanistan: Soviet special operation forces in action." *Military Technology* (9:10) October, 1988, pp. 136-139.

Isley, Jeter, and Philip Crowl. *The U.S. Marines and amphibious war: Its theory and its practice in the Pacific.* Princeton: Princeton University Press, 1951. 636 pp.

Includes extensive references on Evans Carlson, Merritt Edson, the raider battalions, the Raider Regiment, and the 1st and 2nd Parachute Regiments. 1,134 refs.

Jackson, Robert. *The Red Falcons: The Soviet air force in action.* London: Clifton Books, 1970. 236 pp.
References to special units include *Flygflottilj* 19, the Swedish volunteer squadron that served with the Finns; Soviet commandos and saboteurs in the 1939-40 Winter War; the Guards Fighter Regiment; and Soviet paratroops in the Manchurian campaign of 1945.

Jaenecke, Erwin. "Die edelweiss soldaten" [The snow-flower soldiers]. *Deutscher Soldatenkalendar.* Munich: Schild Verlag, 1958, pp. 116-127.
A German account of *Wehrmacht* mountain operations in World War II.

"James Bond? Nein, danke! Die elite-trüppe der polizei mag manner mit angst." [James Bond? No, thanks. The police elite troops prefer men with caution.] *Munchen Merkur*, February 16, 1964, p. 12.

James, Harold, and Denis Sheil-Small. *The Gurkhas.* London: Macdonald, 1965. 291 pp.

James, Malcolm. *Born of the desert.* London: Collins, 1945. 320 pp.
A history of the SAS published at the end of World War II.

Johnson, David, and Phyllis Johnson. *The struggle for Zimbabwe: The Chimurenga war.* London: Faber and Faber, 1981. 378 pp.
Includes accounts of role of Rhodesian *corps d'élite*, raids into Zambia and Mozambique, and tactics. 871 refs.

Johnstone, John H. *United States Marine Corps parachute units.* Washington, D. C.: U.S. Marine Corps, G-3 Division Headquarters, Historical Branch, 1962. 21 pp.
A survey of marine parachute units from conceptualization in early World War II, to formation and ground combat in World War II, through disbandment and re-formation of airborne units with the Force Reconnaissance structure in 1957. 51 refs.

Judge, C.B. "Commandos and their methods." *U.S. Naval Institute Proceedings* (70:7) July, 1944, pp. 843-851.
An overview of British commando operations in World War II from a U.S. perspective, and much filtered by wartime censorship.

Kaltenegger, Roland. *Deutsche gebirgsjäger in zweiten weltkrieg* [German mountain troops in the Second World War]. Stuttgart: Motorbuch Verlag, 1977. 352 pp.
A series of capsule campaign histories focussed at the division level. 82 refs.

Kamenitsky, Ihor. *Secret Nazi plans for Eastern Europe: A study of lebensraum* (living space) *policies*. New York: Bookman Associates, 1961. 263 pp.
Describes in detail the role the *SS* overall played in the Nazi schema for "colonizing" the East, including "Germanization," deportation, extermination, various economic programs, etc. 582 refs.

Keegan, John. *Waffen S.S.: The asphalt soldiers*. New York: Ballantine, 1970. 160 pp.
A brief overview of the *Waffen SS* for the Ballantine paperback history of World War II series. 18 refs.

----------, and Richard Holmes, with John Gay. "Irregulars." In *Soldiers: A history of men in battle*. New York: Viking, 1986, pp. 241-258.
A broad-gauge essay on guerrilla, terrorist, and special operations derived from a BBC TV series. Emphasis is on modern Europe and Chinese Communism, with special attention paid to T.E. Lawrence and Che Guevara, and to special forces-*montagnards* in Vietnam. More impressionistic than definitive.

Kelly, Francis John. *Army special forces, 1961-1971*. Washington, D.C.: U.S. Government Printing Office, 1973. 227 pp.
A straightforward account issued in advance of the official history, with all the problems in respect to sourcing and analysis that such limitations naturally involve. Eases past the crucial question of non-military missions and clandestine functions.

----------. *U.S. army special forces, 1961-1971*. Washington, D.C.: Department of the Army, 1973. 227 pp.
Published immediately after U.S. disengagement in the 2nd Indochina War, this study provides an interim perspective and a useful "Balance Sheet," Chapter VIII.

Kelly, Ned. "Lebanese airborne: On the ground with Lebanon's paras." *Soldier of Fortune* (12:2) February 1987, pp. 32-35.

Kiang, Lindsey (Maj.). "*Fallschirmjäger*: American marine joins German Diving Eagles." *Soldier of Fortune* (9:8) August, 1984, pp. 56-63.

King, Michael J. *Rangers: Selected combat operations in World War II*. Fort Leavenworth: Combat Studies Institute, 1985. 87 pp.

Klenfeld, Gerald R., and Lewis A. Tambs. *Hitler's Spanish legion: The Blue Division in Russia*. Carbondale: Southern Illinois University Press, 1979. 434 pp.

The Blue Division of volunteers was Franco's token assistance to Hitler in the crusade against Bolshevism, far less in the way of cooperation than Hitler hoped or the Allies had feared at the end of the Spanish Civil War. Its volunteers, like those from other Nazi allied and occupied states, suffered heavily from poor equipment, the weather, and the politics of allied relations. 971 refs.

Knight, David C. *Shock troops: The history of elite and special forces*. Greenwich, Conn.: Bison, 1983. 192 pp.

A large format, heavily illustrated, wide ranging survey, popular in style. No refs.

----------. *Shock troops: The history of elite and special forces*. Greenwich, Conn.: Globe, 1983. 192 pp.

A large format, illustrated, popular, selective survey of *corps d'élite*.

Koehl, Robert. *The Black Corps: The structure of the Nazi S.S.* Madison: University of Wisconsin Press, 1983. 437 pp.

A detailed study tracing the complex political and structural aspects of the *SS* system, including links between police, *Einsatzkommando*, and *Waffen SS*, as it evolved into a kind of "department store." The author argues that the *Nur-Soldaten* (soldiers-only) predominated in the *SS*, that only a "fraction of the *Waffen SS* had any connection with the concentration camps," and that the essential elite kernel of the *SS*, its officer corps as it existed in 1943, was decimated in the final phases of the war. 504 refs.

Kohler, David R. "Spetsnaz." *U.S. Naval Institute Proceedings* (113:8) August, 1987, pp. 46-55.

An essay on the Soviet use of commando-type forces under control of the Organs of State Security. Based on English language sources. 41 refs.

Kondrashov, A., and N. Aryayev. "Paratroopers." *Soviet Military Review* (1) January, 1987, pp. 24-27.

Kravtsa, G. P., and G. I. Kornyenko. *Zastaviu y pruta* [Sentry post by the twigs]. Kishineyev: Timpul, 1985. 231 pp.

An illustrated, lyrical view of KGB/MVD border guard forces. Representative of a major effort to dramatize and glamorize *Chekisti*--intelligence operatives, special units, etc.--in Soviet media over the last generation.

Kren, George, and Leon H. Rappoport. "The Waffen SS: A social-psychological perspective." *Armed Forces and Society* (3:1) Winter, 1976, pp. 87-107. 27 refs.

Krepenevich, Andrew F. *The army and Vietnam.* Baltimore: Johns Hopkins, 1986. 318 pp.
Includes extensive discussion and data on special operations doctrine, forces, and operations in the Second Indochina War. 758 refs.

Krulak, Victor H. *First to fight. An inside view of the U.S. Marine Corps.* Annapolis: U.S. Naval Institute, 1984. 384 pp.
A well-illustrated collection of essays on the U.S. Marine Corps by a retired marine general who commanded a parachute battalion in World War II.

Kuhn, Volkmar. *German paratroopers in World War II.* Trans. M. A. and A. J. Barker. London: Ian Allan, 1978. 240 pp.
An overview of German World War II airborne operations. Characterizes Skorzeny's role in the Grand Sasso raid as an upstaging of German airborne forces and General Kurt Student as the main architect of planning and operations. 190 refs.

Kupperman, Robert H., and Darrell M. Trent, eds. *Terrorism: Threat, reality, response.* Stanford: Hoover Institution Press, 1979. 450 pp.
Extensive interthreading of *corps d'élite* and special operations with the complexities of terrorism, both causation and response. Most pertinent is "Specialized Rescue Teams," pp. 153-156.

Ladd, J.D. *Commandos and rangers of World War II.* New York: St. Martin's Press, 1978. 288 pp.

----------. *The invisible raiders: History of the S.B.S. from World War II to the present.* London: Arms and Armour Press, 1983.
A history of the Special Boat Squadron, the Royal Navy's counterpart to the SAS.

Lane, Ronald L. *Rudder's Rangers.* Manassas: Ranger Associates, 1979. 191 pp.
A history of the 2nd Ranger Battalion from formation through the abortive assault on German heavy gun positions on D-Day. 48 refs.

Lebra, Joyce C. *Japanese-trained armies in Southeast Asia.* New York: Columbia University Press, 1977. 226 pp.
While most local forces raised by the Japanese in Southeast Asia were auxiliaries and often unarmed, some were not. Lebra's

detailed analysis includes references to both Japanese and other special units. 187 refs.

Leckie, Robert. *Strong men armed*. New York: Random House, 1962. 563 pp.
Extensive material on U.S. Marine Corps raiders and paramarines. 91 refs.

Lefevre, V. *La division Brandenbourg*. Paris: Presses de la Cité, 1984. 318 pp.
A popular history of the German clandestine force formed in the 1930s, whose main operational principle was deception vs. Allied *corps d'élite* and Skorzeny's emphasis on speed, manuever, and surprise--when their founders and commanders had a choice. 26 refs.

"Legion étrangère, 1831-1981." Special issue of *Revue Historique des Armeés* (1), 1981. 320 pp.
Thirteen articles, mainly by veterans, on battles, campaigns, units, personalities, and the officer and NCO corps of the legion. Includes a chronology and insignia.

Lewy, G. *America in Vietnam*. New York: Oxford University Press, 1978. 540 pp.
A strong but critical *defensa* of U.S. involvement in Vietnam, including details of U.S. and South Vietnamese special operations and *corps d'élite* set in a broader political and operational complex. 1,362 ref.

Lodwick, John. *The filibusters: The story of the special boat service*. London: Methuen, 1947. 189 pp.
A history of SBS World War II operations in North Africa, Italy, the Adriatic, the Balkans, and the Greek Isles.

Lucas, James. *Alpine elite: German mountain troops of World War II*. London: Jane's, 1980. 226 pp.
Substantial detailing of German army and *SS* mountain forces, including order of battle, profiles on principals, and accounts of operations. Also material on uniforms, weapons, special equipment, and tactics. Relatively little on counterguerrilla activities; main emphasis on Poland, Norway, the Balkans campaign, Crete, the Caucasus, and Finland. 32 refs.

----------. *Kommando: German special forces of world war two*. New York: St. Martin's, 1985. 245 pp.
A survey of German World War II land, sea and air special units, focussing mainly on the Brandenburgers, but including details on paratroop special operations and units, KG 200, et al. (Note: the description of Operation Scherhorn is not in agreement with other accounts). 30 refs.

Luttwak, Edward, and Dan Horowitz. *The Israeli army*. New York: Harper and Row, 1975. 461 pp.

Appeared soon after the 1973 October War, and includes background on the evolution of the elements of the army from the Palmach and the paratroops.

Mabire, Jean. *Chasseurs Alpins: Des Vosges aux Djebels, 1914-1964* [Alpine troops: From the Vosges to the Djebels, 1914-1964]. Paris: Presses de la Cité, 1984. 456 pp.

A history of the French army's mountain troops, one of the "*Troupes de choc*" series edited by Jeannine Balland. Much detailing, illustrations. 172 refs.

MacDonald, Charles. *Airborne*. New York: Ballantine, 1970. 160 pp.

A brief survey of airborne operations in World War II. 11 refs.

MacDonald, Peter. *The special forces: A history of the world's elite fighting units*. Secauacus, New Jersey: Chartwell Books, 1987. 256 pp.

A large format, well-illustrated volume. No refs.

----------. *Solders of Fortune: The twentieth century mercenary*. New York: Gallery Books, 1986. 191 pp.

Large format, well-illustrated survey of various regular elite forces including the French Foreign Legion, Gurkhas, the SS, various British imperial forces as well as actual mercenary forces, e.g., the Congo commandos, Rhodesia, Angola. No refs.

MacDougald, N.E. "SOF, special forces & vortex." *Soldier of Fortune* (2:3) September, 1977, 54ff.

"The marine corps." (Royal Dutch). In *Info: Royal Netherlands Navy*. Amsterdam: Information Office of the Ministry of Defense, 1985, pp. 32-33.

Markham, James M. "Commandos of the Soviet bloc said to infiltrate West Europe." *New York Times*, November 2, 1986, Sec. 1, p. 8.

Marks, Tom. "In defense of the realm: Britain's legendary Gurkhas." *Soldier of Fortune* (12:7) July, 1987, pp. 38-45.

Marolda, Edward J., and Oscar P. Fitzgerald. *The United States Navy and the Vietnam conflict. Vol. 2. From military assistance to combat, 1959-1965*. Washington, D.C.: Department of the Navy, Naval Historical Center, 1986. 591 pp.

Extensive material on special operations and SEALs. 130 refs.

McGowan, Robert, Ian Black, Peter Hardy, and Peter Mason. *The day of the SAS*. London: Daily Express Special Edition, 1980. 64 pp.

A detailed, well-illustrated account of the Prince's Gate affair.

McNeill, Ian. *The team: Australian army advisers in Vietnam, 1962-1972*. St. Lucia, Queensland: University of Queensland Press, 1984. 534 pp.

Chapters 12 and 13 describe Australian advisors' 2nd Indochina War service with special forces in I Corps, 1965-1968, and II Corps, 1968-1970.

McNeill, William Hardy. *The Greek dilemma: War and aftermath*. Philadelphia: J.B. Lippincott, 1947. 291 pp.

A first-hand account of the events leading to American involvement in the Greek Civil War. Details on the formation of Greek forces after the 1941 defeat, including the "Sacred Squadron."

Metzger, John. "Battling bellosos: El Salvador's 'tigers' sharpen their claws." *Soldier of Fortune* (8:8) August, 1983, 62ff.

Miksche, F.O. *Paratroops*. London: Faber and Faber, n.d., c. 1942. 167 pp.

Miksche, a Czech exile, became a preeminent military theorist in the 1940s and '50s. The pattern and siting of dropzones on D-Day conformed very closely to Miksche's hypothetical design.

Miles, Milton (Adm.), and Daniel Hawthorne. *A different kind of war: The little known story of the combined guerrilla forces created in China by the U.S. Navy and the Chinese in World War II*. Garden City: Doubleday, 1967. 607 pp.

An account by a principal of the Sino-American Cooperation force--SACO.

Miller, Marshall Lee. "Airborne warfare: A concept the U.S.S.R. actually can claim it invented first." *Armed Forces Journal* (124:3) October, 1986, pp. 48-51.

A brief history of the *Vozdushno Desantnye Voiska*.

Miller, Russell. *The commandos*. New York: Time-Life, 1981. 208 pp.

A well-illustrated volume in the Time-Life World War II series. Uneven and selective, the material on the Marine Raiders being especially thin.

Milliot, Bernard. *Divine thunder*. New York: McCall, 1971. 243 pp.

A very readable account of the special attack corps/ *kamikazes*. 30 refs.

Mollo, Andrew. *To the death's head: The story of the S.S.* London: Thames Methuen, 1982. 131 pp.
A brief paperback history of the *SS* complementing a series on BBC television in the early 1980s. 37 refs.

Morison, Samuel Eliot. *History of United States naval operations in the second world war.* Vol. 14. Boston: Little Brown, 1960. 385 pp.
Includes detailing of underwater demolition units tactics and operations.

Morris, Jim. "Death-dealing project Delta, part 1, Vietnam's delta gave a new meaning to military excellence." *Soldier of Fortune* (6:7) July, 1981, pp. 26-33.

----------. "Death-dealing project Delta, part 2: Author joins Delta on a hot LZ." *Soldier of Fortune* (6:8) August, 1981, pp. 20-25.

----------. "Death-dealing project Delta, part 3: Project commander's military memories." *Soldier of Fortune* (6:9) September, 1981, pp. 47-53.

Motte, Raymond. "Mountain troops in modern warfare." *Military Review* (44:8) August, 1964, pp. 39-45.

Moyer, Frank A., and J. Robert Scorggie. *Special forces combat firing techniques.* Boulder: Paladin Press, 1971. 120 pp.

Moynahan, Brian. "Faithful unto death: The legion's been their country for 150 years." *Soldier of Fortune* (6:12) December, 1981, pp. 44-51.

Musgrove, Gordon. *Pathfinder force: A history of 8 Group.* London: MacDonald and Jane's, 1976. 289 pp.
A history of the Royal Air Force Bomber Command's target-marking force, consisting of the best navigators, guidance technologies, and the first air unit to use the Mosquito bomber, under the command of the dynamic and iconoclastic ex-regular, Don Bennett.

Neild, Eric. *With Pegasus in India: The story of the 153 Gurkha parachute battalion.* N.p.: n.d., c. 1946. 110 pp.

Nelson, Harvey W. *The Chinese military system: An organizational study of the Chinese People's Liberation Army.* Boulder: Westview, 1981. 285 pp.

Traces the effects of the Cultural Revolution on the various services, and aspects of elitism vis-a-vis the airborne and air forces. 106 refs.

Neususs-Hunkel, Ermendhild. *Die S.S.: Schriftenreihe des instituts für wissenschaftliche politik in Marburg/Lahr* [The *SS*: Annals of the Institutes for Political Science in Marburg-Lahu]. Hannover: Norddeutsche Verlags Anstalt, 1956. 143 pp.

Includes analysis of difficulties in determining boundaries between *SS* combat, terror, and control functions.

Never, Reinhard. "Paratroops of the National People's Army: specialists in combat behind enemy lines." *Truppenpraxis*, July, 1983, pp. 585-520.

Newport, Christopher. "Soldiers of the knife. Canadian/ American outfit was too tough for the Wehrmacht." *Soldier of Fortune* (5:7) July, 1980, pp. 64-71.

Nielsen, S. "GSG 9: The unit." *Soldier of Fortune* (5:3) March, 1980, pp. 36-38.

Noonan, William. *The surprising battalion: Australian commandos in China*. Sydney: Bookstall, 1945. 194 pp.

Norton, G.G. *The Red Devils: The story of the British airborne forces*. London: Leo Cooper, 1971. 260 pp.

An unreferenced survey history, in the "Famous Regiment Series," it contains technical appendices, including a detailed table on casualties (Appendix D).

O'Ballance, Edgar. "Who dares wins: SAS from WWII to today." *Soldier of Fortune* (7:12) December, 1982, pp. 80-83.

----------. *Arab guerrilla power, 1967-1972*. Hamden, Conn.: Archon, 1973. 246 pp.

A survey of Arab anti-Israeli guerrilla/commando organization and operations, e.g., *Al-Fatah*.

Office of the Secretary of War, Strategic Services Unit, Historical Project. *War report of the O.S.S.* New York: Walker and Co., 1979. 261 pp.

Basic data on special operations and operational groups' structure, and maritime units.

Ogburn, Charlton, Jr. *The marauders*. New York: Harper and Brothers, 1956. 307 pp.

A history of Merrill's Marauders by a participant.

Olson, James S. *Historical dictionary of the Vietnam War*. Westport, Conn.: Greenwood, 1988. 608 pp.

O'Neill, Richard. *Suicide squads: Axis and Allied special attack weapons of World War II: Their development and their missions*. London: Salamander Books, 1981. 296 pp.

A luridly mistitled survey of British, German, and American special units and operations, with emphasis on the Japanese suicide units. 113 refs., all secondary.

OSS Jedburgh teams I; II. New York: Garland Publishing Co., 1986. 475 p.; 500 p.

Selections from war diaries of the Special Operations Branch of the Office of Strategic Services, the principal American strategic intelligence organization of World War II. Jedburgh teams, each comprised of two or three uniformed members, were dropped into Europe before, on, and after D-Day (1 June 1944) in Operation OVERLORD, the major Allied invasion of northwest Europe. Their main role was coordinating resistance operations with London and arranging supplies.

Other OSS teams. New York: Garland Publishing, 1986. 359 pp.

Excerpts from the war diaries of the Special Operations Branch F Section, originally formed by the British Special Operations Executive, and joined by the American Office of Strategic Services; F Section landed small groups to train, supply, and direct French saboteurs, from mid-1943 to late 1944. In essence, F Section was more of a true antecedent to the U.S. Special Forces raised in the early 1950s than official lineage traced to the 1st U.S.-Canadian Special Service Force and Merrill's Marauders.

Palmer, Bruce, Jr. (Gen.). *The 25-year war: America's military role in Vietnam*. New York: Simon and Schuster, 1984. 236 pp.

Extensive references to U.S. and Vietnamese elite forces and special operations. 122 refs.

Parrah, Shawn M. "Ranger 3000." *New combat systems* (1:3) August, 1987, 46-63.

Popular journalistic view of U.S. rangers in mid-1980s.

Peers, William R., and Dean Brelis. *Behind the Burma road: The story of America's most successful guerrilla forces*. Boston: Little Brown, 1963. 232 pp.

A history of Service Unit Detachment No. 101 of the Office of Strategic Services in Burma in the Second World War, also called the "Kachin Rangers"; includes data on first American airborne commandos--Air Commandos. No refs.

Peniakoff, Vladimir. *Popski's private army*. New York: Thomas Y. Crowell, 1950. 512 pp.

An account of the smallest of the *corps d'élite* to emerge from the North African campaign of World War II, by its founder.

Phillips, C.E. Lucas. *Cockleshell heroes.* London: Heinemann, 1958. 208 pp.

A history of the Royal Marine Boom Patrol, whose exploits were the basis of a feature film.

----------. *The raiders of Arakan.* London: Heinemann, 1971. 198 pp.

A history of V Force, a unit raised from the local population in Burma that performed reconaissance just behind Japanese lines; also includes data on Z Force. Based mainly on interviews and uncited monographs by participants. No refs.

Pigeon, Robert L., Kenneth S. Gallagher, and Edward Wimble. *The soldier of fortune.* New York: Exeter Books, 1986. 184 pp.

A collection of articles from *Soldier of Fortune* magazine including a section on "Elite Men and Elite Units."

Polmar, Norman. "SOF--The navy's perspective." *U.S. Naval Institute Proceedings* (113:8) August, 1987, pp. 136-138.

Includes substantial discussion of equipment and training problems. 2 refs.

Pope, Vernon. "Russian commandos." *Collier's* (112) July 23, 1943, pp. 22-23.

A popular World War II view of Soviet naval infantry.

Powers, W.M. "Almost beyond human endurance." *U.S. Naval Institute Proceedings* (103:12) December, 1977, pp. 60-69.

An illustrated report on the U.S. Navy's Basic Underwater Demolition Team School including SEAL training.

Pratt, Fletcher. *The marines' war.* New York: William Sloane, 1948. 456 pp.

Dated, but very readable, with extensive detailing on the raider battalions.

Pritchard, Charles G. "The Soviet marines." *U.S. Naval Institute Proceedings* (98:3) March, 1972, pp. 18-30.

An article on the *morskaya pekhota* a decade after reconstitution. Soviet naval infantry, highly visible in World War II, had apparently been deactivated 1947-1963.

Quarrie, Bruce. *The world's elite forces.* New York: Berkley Books, 1985. 248 pp.

A mid-1980's survey of the elite forces of Britain and the Commonwealth, the U.S., Western Europe, Israel and the Warsaw Pact. No refs.

----------. *The world's secret police.* London: Octopus Books, 1986. 160 pp.

A brief survey of security forces, with extensive detailing of human rights abuses, drawing heavily on Amnesty International reports; uneven in respect to data on organizations.

Rademaker, John A. *These are Americans: The Japanese Americans in Hawaii in World War II*. Palo Alto: Pacific Books, 1951. 278 pp.
Extensive detailing of origins, training, and combat actions of Japanese-American 442nd Regimental Combat Team and the 100th Battalion, including maps and statistics.

R.A.F. airborne forces manual. London: Arms and Armour Press, 1979. 288 pp.
Reproductions of World War II documents relating to British airborne forces.

Rapaport, Leonard, and Arthur Norwood, Jr. *Rendezvous with destiny: The story of the 101st*. Washington: Infantry Journal Press, 1948. 810 pp.
A detailed, adulatory history of the U.S. 101st Airborne Infantry Division in World War II, including honor roll, statistics, and songs.

Ravenstein, Charles A. *Air force combat wings: Lineage and honors histories, 1947-1977*. Washington, D.C.: United States Air Force, Office of Air Force History, 1984. 341 pp.
Includes data on special operations units dating from the Air Commandos of World War II.

Reed, Fred. "U.S. Ranger school: Instead of 3 Rs, Rangers learn to patrol, patrol and patrol." *Soldier of Fortune* (6:10) October, 1981, pp. 52-57.

Reid-Daly, Ron, and Peter Stiff. *Selous scouts: Top secret war*. Alberton: Galago Publishing Co., 1982. 432 pp.
A lengthy and well-illustrated account by its founder of the most controversial unit in the array of Rhodesian special units during the operations that led to the end of white settler rule and the creation of Zimbabwe, 1973-1980. 27 refs.

Reitlinger, Gerald. *The SS: Alibi of a nation, 1922-1945*. London: Heinemann, 1956. 502 pp.
A careful post-mortem of the *SS* as an institution, the blurred roles and unclear motives and doctrine. 117 refs.

Riad, el-Rayyes, and Dunia Nahas. *Guerrilla for Palestine*. London: Croom Helm, 1976. 155 pp.
Contains substantial data on lineage of various anti-Israeli and Palestinian commando units from the 1930s through the period immediately following the Munich Olympic massacre and as Israeli retaliation was in process. Includes principal leaders, background on *Al-Fatah*, Black September, et al.

Ritter, J.J. "UDT comes of age." *U.S. Naval Institute Proceedings* (91:2) February, 1965, pp. 52-63.
A view of U.S. Navy underwater special operations units on the eve of major expansion and commitment in the 2nd Indochina War, and three years after the creation of the SEALs.

Roggiani, Fermo. *Storia del Bersaglieri d'Italia.* Milano: Canlotti Editori, 1973. 453 pp.
A history of the Italian army light infantry corps.

Rolbant, Samuel. *The Israeli soldiers: Profile of an army.* New York: Thomas Yozeloff, 1970. 353 pp.

Roscoe, Theodore. *United States submarine operations in World War II.* Annapolis: U.S. Naval Institute, 1949. 577 pp.
Although dated in sourcing and some data, it still stands as the best illustrated work on the subject, and includes much data on special operations.

Ross, Jane. "Australian soldiers in Vietnam: Presence and performance." In *Australia's Vietnam: Australians in the Second Indochina War*, ed. Peter King. Sydney: George Allen and Unwin, 1983, pp. 72-99.
Includes information on the role of the Australian SAS in Vietnam. 111 refs.

Rothenberg, Gunther. *The anatomy of the Israeli army.* New York: Hippocrene Books, 1979. 256 pp.
Includes substantial detail on Israeli army *corps d'élite.* 180 refs.

Ryan, Paul B. *The Iranian rescue mission: Why it failed.* Annapolis: U.S. Naval Institute Proceedings, 1985. 136 pp.
A post-mortem of Operation *Eagle Claw.*

Salisbury, Steve. "Mission misura: Cross-border ops with contras special forces." *Soldier of Fortune* (9:11) November, 1984, pp. 80-87.

Saunders, Hilary St. George. *Combined operations: The official story of the commandos.* New York: Macmillan, 1943. 155 pp.

----------. *The Green Beret: The story of the commandos, 1940-1945.* London: Michael Joseph, 1949. 362 pp.
The first major post-World War II history of the commandos. No refs.

Scaramanga, Philip K. "The Spanish legion: Professional core of Spain's rapid deployment force." *International Defense Review* (21:3), 1988, pp. 273-274.

A detailed description of the structure and current assignments of the Spanish Foreign Legion in the wake and prospect of reorganization and reassignment.

Schauer, Hartmut. *Die geschichte einer elitetruppe.* Stuttgart: Motorbuch Verlag, 1987. 210 pp.

Schiff, Zeev, and Raphael Rothstein. *Fedayeen: Guerrillas against Israel.* New York: Mckay, 1972. 246 pp.

Schnabel, Reimund E., ed. *Macht ohne moral: Eine dokumentation über die SS* [Power without morality: A documentation of the *SS*]. Frankfurt-am-Main: Roderberg Verlag, 1957. 567 pp. 107 refs.

Schofield, William, and P.J. Carisella. *Frogman first battles.* Brookline Village, Mass.: Branden, 1987.
A history of the "Tenth Light Flotilla".

Scurr, John. *The Spanish foreign legion.* London: Osprey Publishing, 1985. 48 pp.
A well-illustrated account drawn from Spanish and English sources. 8 refs.

Sergeev, S.S. *Totalny Shpionazh.* Moscow: Voenizdat, 1984.
Discussion of Western "special operations" against socialist forces, pp. 140-146.

Sergent, Pierre. *Histoire mondiale des parachutistes.* Paris: Société de Production Littéraire, 1974. 368 pp.

Seth, Ronald. *Lion and blue wings: The story of the glider pilot regiment, 1942-1945.* London: Victor Gollancz, 1955. 245 pp.

Seymour, William. *British special forces: The story of Britain's undercover soldiers.* London: Grafton Books, 1985. 512 pp.
A survey of British special units, from the 1930s to the 1980s. 129 refs.

Shaw, W.B.K. *Long range desert group: The story of its work in Libya, 1940-1943.* London: Collins, 1945. 256 pp. (2nd ed., London: Landsborough Publications, 1959).

Sherrod, Robert. *History of marine corps aviation in World War II.* San Rafael: Presidio Press, 1980. 496 pp.
A reprint of World War II Time correspondent Sherrod's well-sourced treatise on Marine Corps air elements. Approx. 400 refs.

Shirley, Orville C. *Americans: The story of the 442nd combat team*. Washington: Infantry Journal Press. 151 pp.
A history of a Japanese-American regimental combat team, the most decorated of its size in the Italian-European campaigns, raised from volunteers among the detention camps and the Japanese-American population in World War II, but often omitted from the litany of American *corps d'élite*.

Shortt, Jim. "Rebuffing the bear." *Soldier of Fortune* (12:12) December, 1987, pp. 84-87.

Showalter, Dennis. "Evolution of the U.S. Marine Corps as a military elite." *Marine Corps Gazette* (63:11) November, 1979, pp. 44-57.
Showalter examines the tension in the history of the U.S. Marine Corps between roles and self-images, and argues that its elite status has been maintained, in spite of the decline in actual functions, and in traditional standards of voluntarism and quality in the enlisted force in Vietnam. The essence of his analysis is that "images and realities conflict sharply." It also touches on the larger questions of perception and definition.

Simpson, Charles H., III. *Inside the Green Berets: The first thirty years*. Novato: Presidio Press, 1983.
An impassioned and well-documented brief for special operations, including a refutation of the argument of Slim et al. that *corps d'élite* drain scarce leaders from the general force. Also includes critiques of CIA misuses, publicity, and calls for a shift to use as a "force multiplier" as deep penetration harassment forces against a major Warsaw pact assault.

Simpson, Howard. "Sappers of the legion: The 6th regiment of the French Foreign Legion." *Soldier of Fortune* (10:12) December, 1985, pp. 48-51.

----------. "Commando: Royal marines hit the beach in Norway." *Soldier of Fortune* (12:7) July, 1987, pp. 56-61.

Skorzeny, Otto. *Meine kommando unternehmen* [My commando operations]. Wiesbaden/ Munich: Limes Verlag, 1976. 445 pp.
A memoir of his World War II exploits by the pre-eminent Nazi practitioner of special operations.

Smith, Cecil B., Jr. *Special forces in Southeast Asia*. Erin, Tenn.: ARV-CAT, 1978. 44 pp.
A list of U.S. and South Vietnamese special and paramilitary units of the 2nd Indochina War, and insignia. No refs.

Smith, R. Harris. *O.S.S.: The secret history of America's first intelligence agency*. Berkeley: University of California Press, 1981. 458 pp.

The definitive history to date of the CIA's predecessor. Expands on details of structure and operations laid forth in the War Report of the OSS. 839 refs.

SOF Staff. "Germany's winter warriors: Mountain troops rise to new heights." *Soldier of Fortune* (9:10) October, 1984, pp. 54-55.

Sorge, Martin K. "The fate of special forces." In *The other price of Hitler's war: German military and civilian losses resulting from World War II*. Westport, Conn.: Greenwood, 1986, pp. 42-54.

Chapter 4 focuses on casualties in *Waffen SS*, airborne, mountain troops and other, non-elite forces; data presentation is uneven, e.g., "losses" sometimes include killed, wounded, and missing, other times killed only; Skorzeny is not mentioned. The bibliography is the most useful element. 376 refs.

Spaeter, Helmuth, *Die Brandenburger--eine Deutsche kommandotruppe: Lehr Regiment ZV800* [The Brandenburger--A German commando force: Training Regiment ZV800]. Munich: Verlagsagentum Walther Angerer, 1978. 544 pp.

A detailed history of the special deception-penetration units of the German *Abwehr*, including details of organization, training, and operations, with many maps. 221 refs.

Special forces demolition techniques. Boulder: Paladin Press, n.d., c. 1970. 67 pp.

"Special units get high-level look." *U.S. Naval Institute Proceedings* (94:8) August, 1968, p. 152.

A glimpse at some of the state-of-the-art underwater special warfare techniques at height of U.S. operations in Vietnam in the 2nd Indochina War.

Spector, Ronald. *Advice and support: The early years*. Washington, D.C.: Center for Military History, 1983. 391 pp.

The first volume of the "U.S. Army in Vietnam" official history, sourced from official documents and detailing the debates and convolutions of U.S. policy in Indochina from 1945 to 1961. Includes material on Brink and Lansdale, and U.S. links with the French.

Stanhope, Henry. *The soldiers: An anatomy of the British army*. London: Hamish Hamilton, 1979. 372 pp.

A survey by a senior British defense journalist including chapters on: "Corps d'Élite," encompassing the Household Division, Greenjackets, and Royal Horse Artillery; "The SAS and the Paras"; and "The Gurkhas." No refs.

Stanton, Shelby L. *Green Berets at war: U.S. Army special forces in Southeast Asia, 1956-1975*. London: Arms and Armour Press, 1986. 360 pp.

The most detailed account extant on a subject upon which the official history may ultimately shed further light. 256 refs.

Stein, George. *The Waffen S.S.: Hitler's elite guard at war, 1939-1945*. Ithaca: Cornell University Press, 1966. 313 pp.

Focuses on the combat role of the armed *SS*. 125 refs.

Stevenson, William. *Zanek! A chronicle of the Israeli air force*. New York: Viking, 1971. 344 pp.

A popular view of the Israeli air force, the central *corps d'élite* in the Israeli defense force.

Strawson, John. *A history of the SAS regiment*. Secker and Warburg, 1984. 292 pp. No refs.

Strutton, Bill, and Michael Pearson. *The secret invaders*. London: Hodder and Stoughton, 1958. 287 pp.

An account of C.O.P.P.--Combined Operations Pilotage Parties--in the Second World War.

Summers, Harry G. *Vietnam war almanac*. New York: Facts on File Publication, 1985. 414 pp.

Extensive details on special forces and operations. Approx. 400 refs.

Suvorov, Viktor. *Inside the aquarium: The making of a top Soviet spy*. New York: Macmillan, 1986. 249 pp.

----------. *Inside Soviet military intelligence*. New York: Macmillan, 1984. 193 pp.

Suvorov, a defector from Soviet GRU--military intelligence--and a veteran of *spetsnaz* service, describes the recruitment, training, security procedures, and operational methods of the *spetsnaz*.

----------. "Spetsnaz: The Soviet Union's special forces." *Military Review* (54:3) March, 1984, pp. 30-46.

Swiggett. *March or die*. New York: Putnam, 1953. 239 pp.

Dated and unreferenced but not without interest, especially Chapter 21, pp. 214-226, "The Legionnaire in Fiction."

Swinson, Arthur. *The raiders: Desert strike force*. New York: Ballantine, 1968. 160 pp.

A brief overview of "mobs for jobs," the small raiding and reconaissance forces in North Africa, 1941-43. 8 refs.

Sydnor, Charles W. *Soldiers of destruction: The S.S. death's head division, 1933-1945*. Princeton: Princeton University, 1977. 371 pp.

A carefully researched tracing of cross-linkages between the *Waffen SS* and the concentration camp system, which found rather more cross-movement than suggested by defenders of the former. Bibliography includes *SS* document ranges in the Bundesarkhiv, Koblenz, FRG, and in the National Archives in Washington. 233 refs.

Sykes, Christopher. *Orde Wingate*. Cleveland: World Publishing, 1959. 576 pp.

A highly sympathetic biography of Wingate, based on family papers, and thinly sourced and referenced.

Talbott, John E. *The war without a name: France in Algeria, 1954-1962*. New York: Knopf and Random House, 1980. 305 pp.

An overview of the war that generated the revolt of the *paras*, the coming to power of de Gaulle, and the Secret Army Organization.

Taubman, Philip. "The secret world of a Green Beret." *New York Times Magazine*, July 4, 1982, pp. 18, 19, 21, 22, and 24.

An account by Special Forces veteran on the purported covert training of Libyans, 1977-78.

Taylor, Chuck. "GSG 9: The guns." *Soldier of Fortune* (5:3) March, 1980, pp. 39-42.

Templeton, Kenneth S. *10th mountain division: America's ski troops*. Chicago: privately published, 1945. 108 pp.

An introductory essay, letters, portraits of members of the 10th, and operational photographs. No refs.

Terblanche, Stephan. "RECCE commando: Stalking prey in the southern African bush." *Soldier of Fortune* (11:1) January, 1986, pp. 76-83.

Thomas, Lowell. *Back to Mandalay*. New York: Greystone Press, 1951. 321 pp.

A well-illustrated popular history of the origins of the "Air Commandos"--special operations groups--under Phil Cochran and John Alison in Burma in World War II.

Thompson, Leroy. "*Cazadores españoles*: Spain's special forces." *Soldier of Fortune* (10:8) August, 1985, pp. 38-39.

----------. "Islamic special forces: Muslim military capabilities." *Soldier of Fortune* (11:4) April, 1986, pp. 34-37.

----------. "SAS today: Tough training for tough soldiers." *Soldier of Fortune* (7:12) December, 1982, 84ff.

----------. "Swiss shock troops: Elite warriors guard fortress Switzerland." *Soldier of Fortune* (9:9) September, 1984, pp. 46-47.

----------. *U.S. elite forces--Vietnam.* Carrolltown, Texas: Squadron/Signal Publications, 1985. 49 pp.
A well-illustrated short survey of U.S. special operations and units in the 2nd Indochina War.

Tickell, Jerrard. *Moon squadron.* Garden City: Doubleday, 1958. 204 pp.
A history of the activities of 138 and 161 squadrons of the Royal Air Force, who supported special operations. The main focus is northwest Europe, with a final chapter on the Far East. No refs.

Tinnin, David B., and David Halevy. "Strike teams." *Playboy* (26:2) February, 1979, pp. 90-92, 94, and 191-197.
An article on SAS, GSG 9, Sayaret Matkal, Blue Light, and Rangers, including scenarios of response to aircraft hijacking, and brief descriptions of special counterterror operations at Mogadishu, Lanarca, Entebbe, and in the retaliation for Lod.

Tophoven, Rolf. *GSG-9: Kommando gegen terrorismus* [Commando against terrorism]. Koblenz: Weher and Wissen, 1977. 96 pp.
A popular format, well-illustrated volume, including data on origins, weapons, training, and operations of *Grenzschützgruppe 9* (Frontier Security Group 9), the Federal Republic of Germany's antiterror force. No refs.

Tranoy, Arnstein. "Norway's cold weather commandos: Preparing for war in the fjords." *Soldier of Fortune* (11:12) December, 1986, pp. 38-43.

Tugwell, Maurice. *Airborne to battle: A history of airborne warfare, 1918-1971.* London: William Kimber, 1971. 367 pp.
Perhaps the best one-volume history of airborne operations; including analysis. 95 refs.

Tuker, Francis. (Lt. Gen. Sir). "Gurkha riflemen." In *Gurkha: The story of the Gurkhas of Nepal.* London: Constable and Co., 1957, pp. 217-230.
A brief description of World War II operations of a primary *corps d'élite* of the old and new British and Indian armies, when the 10 battalions were increased to 45. Includes a bitter condemnation of Orde Wingate's long-range penetration concept, and criticism for his mishandling of Gurkha elements. 47 refs.

Turbiville, Graham H. "Soviet airborne forces." *Army* (26:4) April, 1976, pp. 18-22, 24 and 26-27.

Uhlig, Frank. *Vietnam: The naval story*. Annapolis: U.S. Naval Institute Proceedings, 1986. 515 pp.
Includes many references on American and South Vietnamese special units in the 2nd Indochina War.

Updegraph, Charles L. *U.S. Marine Corps special units of World War II*. Washington, D.C.: U.S. Marine Corps Headquarters Historical Division, 1977. 105 pp. 63 refs.

Vagts, Alfred. *Hitler's second army*. Washington, D.C.: Infantry Journal Press, 1943. 245 pp.

Van Atta, Dale. "Spetsnaz: The Soviets' sinister strike forces." *Reader's Digest* (128:768) April, 1986, pp. 72-76.

Venter, Al J. "French reaction force: Keeping the peace in Africa." *Soldier of Fortune* (12:1) January, 1987, pp. 44-47.

----------. "RECCE commandos--Rugged training for rugged troops." *Soldier of Fortune* (3:4), July, 1978, pp. 39-43.

Waldron, T. J., and James Gleeson. *The frogmen: The story of the wartime underwater operators*. London: Evans Brothers, 1954. 191 pp.
A popular history of mainly British World War II underwater operations, including Italian origins, based on a BBC production by the authors. No refs.

Walmer, Max. *An illustrated guide to modern elite forces*. New York: Arco Publishing, 1984. 159 pp.
A graphically slick, unreferenced handbook of 30 *corps d'élite* of all continents except Antarctica and South America.

Warman, Mark. "Afghan elite forces: Mujahideen factions join forces." *Soldier of Fortune* (8:12) December, 1983, pp. 62-71.

Warner Philip. *The special air service*. London: William Kimber, 1980. 280 pp.
A detailed unreferenced history of the SAS, 1941-1964, with seven appendices.

Warren, C.E.T., and James Benson. *The midget raiders: The wartime story of human torpedoes and midget submarines*. New York: William Sloane, 1954. 318 pp.
Based on extensive interviews with veterans of the Royal Navy's World War II special underwater operations.

Warren, John C. *Airborne missions in the Mediterranean, 1942-1945*. Manhattan, Ks.: Sunflower Press, c. 1985. [Reprint of U.S. Air Force Historical Studies No. 74 (1953).] 137 pp.

----------. *Airborne operations in world war II*. European theater. Manhattan, Ks.: Sunflower Press, c. 1985. [Reprint of U.S. Air Force Historical Studies No. 97 (1956).] 293 pp.

Watts, D.H., and J.J. Ritter. "Combat swimmers." *U.S. Naval Institute Proceedings* (91:5) May, 1965, pp. 95-111.
A pictorial feature showing hydrographic reconnaissance, demolitions, recovery techniques, miniature craft, and support systems of underwater demolition teams and SEALs.

Weingartner, James J. *Hitler's guard: The story of the Leibstandarte S.S. Adolf Hitler, 1933-1945*. Carbondale: Southern Illinois University Press, 1968. 194 pp.
A "study in Nazi feudalism," the history of a bodyguard regiment which was a central *corps d'élite* within the *SS Verfügungstruppe* before World War II, and which was ultimately expanded into a panzer (armored) division. 94 refs.

Wellard, James. *The French Foreign Legion*. Boston: Little Brown, 1974. 142 pp.
A brief, popular, well-illustrated history of the legion. 53 refs.

"Whatever happened to - Green Berets: Peace specialists now." *U.S. News and World Report*. November 6, 1972.

White, David Fairbank. "SEALs are the Navy's elite amphibious force: Are they the toughest men alive?" *Parade*, August 16, 1987, pp. 4-5.

Whiting, Charles. *Hunters from the sky: The history of the German parachute regiment, 1940-1945*. London: Leo Cooper, 1974. 209 pp.

Wilmot, Chester. *The struggle for Europe*. New York: Harper and Brothers, 1952. 766 pp.
Wilmot's tour de force remains a standard source, although written from sources available just after World War II. Wilmot, an Australian war correspondent, jumped into Normandy with British paratroops on D-Day. Includes detail on airborne forces, such as the Germans' use of the airborne mystique to motivate young soldiers who had never jumped, and on other elite forces, e.g., the 79th Armoured Division. 94 refs.

Wright, Bruce S. *The frogmen of Burma: The story of the sea reconnaissance unit*. Toronto: Clarke, Irwin. 152 pp.
A popular account of a Royal Navy swimming unit trained in the Bahamas and California in World War II, that served as a

reconnaissance and guide force in Burma. Includes details on training, and on other underwater units including Japanese swimmer units.

Young, B.A. *The artists and the SAS: London 21st Special Air Service regiment (Artists Rifles)*. London: Territorial Army, 1960. 63 pp.

3
Special Operations in Major Wars

Adleman, Robert H. *The Champagne campaign.* Boston: Little Brown, 1969. 298 pp.

A survey of Operation ANVIL-DRAGOON, the invasion of Southern France in August 1944, including commando, airborne, and 1st U.S.-Canadian Special Service Force involvement. 310 refs.

Aerospace Studies Institute, Concepts Division. *Guerrilla warfare and airpower in Korea, 1950-53.* Maxwell Air Force Base: Air University, 1964. 248 pp.

Includes details on special operations and units in Korea, including North Korean special operations; U.S. Marine recon operations; Korean (South) National Police combat teams; partisan liaison and various U.S. elements; insertions; psychological warfare; logistics of partisan supply; and personnel and organizational problems encountered. 181 refs.

Ambler, John Stewart. *The French army in politics, 1945-1962.* Bowling Green: Ohio State University Press, 1966. 427 pp.

Substantial discussion and references to *corps d'élite* and of political and psychological dimensions of special operations in French colonial wars. 481 refs.

Anthony, Arthur. *Deliverance at Los Baños.* New York: St. Martin's, c. 1985. 285 pp.

An account of the U.S. airborne raid which freed U.S. prisoners at the Los Baños internment camp in the Philippines in February 1945. 109 refs.

Aristu, Captain Javier. "Argentine elite: Special ops in the Falklands." *Soldier of Fortune* (10:12) December, 1985, pp. 76-81.

Arnold, Henry H. "The aerial invasion of Burma." *National Geographic Magazine* (86:8) August, 1944, pp. 129-148.

Atkin, Ronald. *Dieppe 1942: The jubilee disaster.* London: Macmillan, 1980. 306 pp.

A popular history of the ill-fated major raid against the toughest portion of German coastal defenses in Europe in August 1942 by a mainly Canadian force. Commandos also played a role and 50 U.S. Rangers accompanied them. Some have judged Dieppe as a planned failure to quieten Soviet demands for a "Second Front" or a political ploy forced by Canadian forces in Britain being inactive since 1939. The author, however, sees a positive effect on the D-Day landing.

Baron, Richard W., Abraham J. Baum, and Richard Goldhurst. *Raid! The untold story of Patton's secret mission.* New York: G.P. Putnams, 1981. 212 pp.

Based on interviews of veterans of the Hammelburg raid, in which an armored task force was ordered by U.S. Third Army commander General George Patton to strike fifty miles inside German lines to liberate a German prisoner-of-war camp in which Patton's son-in-law was held. Task Force Baum was unsuccessful, losing all its vehicles and suffering over twenty percent casualties in a highly controversial exercise of lèse-majesté, and is an instance of a special operation assigned to a standard force.

Bergôt, Erwan. *Le dossiėr rouge: Services secret contre FLN.* Paris: Bernard Grasset, 1976. 314 pp.

Presents details regarding French covert/special operations against the Algerian nationalists, from the early 1950s to French withdrawal.

Bidwell, Shelford. *The Chindit war: Stilwell, Wingate and the campaign in Burma, 1944.* New York: Macmillan Publishing Co., 1979. 304 pp.

A very readable and personal account of events in Burma, political and military, by a retired senior British officer known for his incisive style and iconoclasms.

Brou, Willy. *Combat beneath the sea.* Trans. Edward Fitzgerald. New York: Crowell, 1957. 240 pp.

A survey of U.S., British, German, and Italian World War II underwater special operations. No refs.

Chandler, Robert W. *War of ideas: The United States campaign in Vietnam.* Boulder: Westview, 1981. 301 pp.

Focuses on various programs, e.g., *Chieu Hoi*, "Hearts and Minds," and propaganda in North Vietnam; concluding section on "Lessons Learned." 160 refs.

Clutterbuck, Richard L. *The long, long war: Counter-insurgency in Malaya and Vietnam.* New York: Praeger, 1966. 206 pp.

Describes SAS role, and special operations and psychological warfare aspects. 21 refs.

Cruickshank, Charles G. *Deception in World War II.* Oxford: Oxford University Press, 1979. 248 pp.

A study of the "London Control Section" which oversaw deception efforts covering major invasions of Western Europe. Primarily British in focus; Germans, Russians, Italians, etc. are dealt with *en passant*. 460 refs.

----------. *The fourth arm: Psychological warfare, 1938-1945*. London: Davis-Poynter, 1977. 190 pp. 63 refs.

David, Heather. *Operation: Rescue--The first complete account of the daring coup into North Vietnam to rescue American prisoners of war*. New York: Pinnacle Books, 1971.

de la Gorce, Paul-Marie. *The French army*. Trans. Kenneth Douglas. New York: Braziller, 1963. 568 pp.

A standard history of the modern French army, completed as de Gaulle consolidated his power; focuses closely on the *paras* in Indochina and Algeria.

de la Penne, Luigi, and Virgilio, Spigai. "The Italian attack on the Alexandria naval base." *U.S. Naval Institute Proceedings* (82:2) February 1956, pp. 125-135.

Description of preparations for and attacks that sank four British warships and 26 merchantmen in World War II.

Devins, Joseph H. *The Väagso raid: The commando attack that changed the course of World War II*. Philadelphia: Chilton, 1968. 222 pp.

A detailed account and extensive background on the Väagso raid, claiming its impact on Hitler diverted major German resources to Norway, and on the Allied side, strengthened joint and inter-Allied liaison.

Doyle, Chris. "Delta ambush. Kit Carson scouts: An action arm of the Phoenix program." *Soldier of Fortune* (5:6) June, 1980, pp. 40-45.

Drummond, John D. *But for these men: How eleven commandos saved Western civilization*. London: Allen, 1962. 205 pp.

Dwyer, John B. "UDT's in Korea: Underwater warriors at war." *Soldier of Fortune* (11:9) September, 1986, 70ff.

Embry, Basil (Air Chief Marshal Sir). "D-Day--Mosquitos foil the Gestapo." In *Bombs away: True stories of strategic air power, from World War I to the present*, ed. Stanley M. Ulanoff. Garden City: Doubleday, 1971, pp. 277-298.

Includes planning, execution, and effects of major low-level attacks by RAF medium bombers against Nazi key point targets in France, Holland, and Denmark, 1944-1945.

----------. "The raid on Shell House." In *War in the air: The Royal Air Force in World War II*, ed. Gavin Lyall. New York: Morrow, 1969, pp. 360-363.

An account of the pinpoint attack on a Gestapo records center in The Hague.

Flanagan, Edward M. *The Los Baños raid: The 11th Airborne jumps at dawn.* Novato: Presidio Press, 1986. 276 pp. An account of the Los Baños raid by a veteran of the raid, later a general officer. 91 refs.

Frere-Cook, Gervis. *The attacks on the Tirpitz.* Annapolis: U.S. Naval Institute Press, 1973. 112 pp.

Gallagher, Thomas. *Assault in Norway: Sabotaging the Nazi nuclear bomb.* New York: Harcourt Brace Javonovich, 1975. 234 pp.

An account of a raid on the Vermork plant in October 1942 by Norwegians trained by SOE, based on interviews, on-the-ground study, and research in the British Public Record Office and the National Archives. No refs.

Geer, Andrew Clare. *The new breed: The story of the U.S. Marines in Korea.* New York: Harper and Brothers, 1952. 395 pp.

Includes references to reconnaissance companies, 41st Royal Marine Commando, underwater demolition teams, and the Special Operations Group.

Golley, John. *The big drop: The guns of Merville, June 1944.* London: Jane's Publishing Co., 1982. 174 pp.

A brief history of a special operation aimed at removing the threat of a German artillery battery to the flank of the British 6th airborne division during the Normandy landing, 1944. 15 refs.

Graves, Jim, and Bob Poos. "The empire strikes back: Royal marine SBS and SAS show their stuff in the Falklands." *Soldier of Fortune* (7:12) December, 1982, pp. 74-79.

Hamilton, Alex. *Wings of the night: The secret missions of Group-Captain Charles Pickard.* London: William Kimber, 1977. 175 pp.

One of the few accounts of air forces in support of clandestine/special operations.

Hampshire, A. Cecil. "The exploits of 'Force Viper.'" *Journal of the Royal United Services Institute* (63:1) February, 1968, pp. 41-50.

An account of commando operations in Burma in World War II.

Haukeld, Knut. *Attack on Telemark.* New York: Ballantine, 1974. 160 pp.

----------. *Skis against the atom.* New York: Ryerson, 1954. 201 pp.

Both are accounts of raids against a facility in Norway in World War II suspected at the time to be part of a major Nazi effort to build an atomic bomb.

Heggoy, Alf Andrew. *Insurgency and counterinsurgency in Algeria.* Bloomington: Indiana University Press, 1972. 327 pp. 420 refs.

Hoettl, Wilhelm. *Secret front: The story of Nazi political espionage*, trans. R.H. Stevens. New York: Praeger, 1954. 327 pp.

Sets Skorzeny's activities in the context of the overall operational perspective of the *SS Sicherheitdienst* (Security Service) and *Wehrmacht* intelligence, by a veteran of Branch VI, under Walter Schellenberger.

Horan, H.E. "'Operation Chariot': The raid on St. Nazaire, 27th-28th March, 1942." *Journal of the Royal United Service Institution* (106:5) February, 1961, pp. 561-566.

Horne, Alistair. *A savage war of peace: Algeria 1954-1962.* New York: Viking Press, 1978.

Extensive detailing on French elite forces and special operations, including the SDECE and the 11th Shock. 186 refs.

Howarth, Patrick, ed. *Special operations.* London: Routledge and Kegan Paul, 1955. 239 pp.

Hughes-Hallett, J. "The mounting of raids." *Journal of the Royal United Services Institution* (95:5) February-November, 1950, pp. 580-588.

Jones, R.V. *The wizard war: British scientific intelligence, 1939-1945.* New York: Coward, McCann and Geoghagen, 1978. 556 pp.

A participant scientist's lucid account of the development of various electronic warfare systems in World War II, offering detailed perspectives on links between technical intelligence and special operations, e.g., the Bruneval raid. 31 refs.

Karig, Walter. "The Makin Island raid." *U.S. Naval Institute Proceedings* (72:10) October, 1946, pp. 1277-1282.

An enthusiastic and uncritical account, claiming as gains the diversions of a Japanese task force from the Solomons, and familiarity gained with Japanese weapons and practices as offsetting the confusion and relatively high casualties. No refs.

Kelly, George Armstrong. *Lost soldiers: The French army and empire in crisis, 1947-1962*. Cambridge: MIT Press, 1965. 404 pp.

A detailed exposition of the complex interactions between *corps d'élite*, special operations, psychological warfare, and politics. 163 refs.

King, Michael J. *Rangers: Selected combat operations in World War II*. Fort Leavenworth: Combat Studies Institute, 1986. [Leavenworth Pamphlet No. 11.] 88 pp.

An analysis of the origins and training of the Rangers, and operations at Djebel el Ank, Porta Empedocle, Cisterna, Zerf, and Cabanatuan; many archival source documents are identified in the bibliography. 168 refs.

Landsborough, Gordon. *Tobruk commando*. London: Cassell, 1956. 216 pp.

A history, in popular style, of the ill-fated raid by commandos, the Long-Range Desert Group, SAS, and SIG elements against Tobruk, with Royal naval support in August-September 1942. No refs.

Lepotier, Adolphe Auguste Marie. *Raiders from the sea*. London: Kimber, 1954. 200 pp.

A French account of the raids on St. Nazaire and Dieppe.

Lerner, Daniel. *Sykewar: Psychological warfare against Germany, D-Day to V-E Day*. New York: George W. Stewart, 1979. 463 pp.

Based on the exhaustive body of material available at the time. 1,340 refs.

Lorain, Pierre. *Clandestine operations: The arms and techniques of the resistance, 1941-1944*, ed. David Kahn. New York: Macmillan, 1983. 185 pp.

A compilation of technical data on radios, radio detection, cryptography, aircraft, organizational structure and process, weapons and explosives, and appendices including tabular and graphic data.

Loustan, Henry-Jean. *Guerre en Kabyli, 1956-1961*. 248 pp.

A history of the field operations of the French army in the mountains of Algeria, including the development of heliborne tactics which relied on the *paras* as the key shock element.

Lovell, Stanley P. *Of spies and strategems*. New York: Pocket Books, 1964. 229 pp.

An account of the author's experiences in special operations; related research and development in the Office of Strategic Service in World War II.

Macksey, Kenneth J. *The partisans of Europe in the Second World War.* New York: Stein and Day, 1975. 258 pp.
Includes references to various special operations and units.

Margolin, Leo J. *Paper bullets: A brief history of psychological warfare in World War II.* New York: Froben Press, 1946. 149 pp.
Emphasis is on leaflets, of which several facsimiles are included.

Marshall, S.L.A. *Ambush.* New York: Cowles, 1969. 242 pp.
A compendium of unit actions in Vietnam, including a chapter on Special Forces' Operation Blackjack 22.

----------. *Night drop: The American airborne invasion of Normandy.* Boston: Little Brown, 1962. 425 pp.
A classic account by the dean of American World War II military historians, based heavily on post-action interviews close to events.

Mason, David. *Raid on St. Nazaire.* New York: Ballantine Books, 1970. 157 pp.
A paperback volume of the Ballantine World War II series.

Maund, L.E.H. *Assault from the sea.* London: Methuen, 1949. 311 pp.

McKie, Ronald. *The heroes.* New York: Harcourt-Brace, 1960. 307 pp.
A description of two light-craft raids by the Australian Z-Force against Japanese bases in Southeast Asia in World War II; Operation Jaywick in 1943, and Operation Rimau (tiger) in 1944. The raiders in the latter were captured and executed. The author concludes with a critique of Charles Willoughby's (MacArthur's chief-of-staff for intelligence) characterization of the missions as suicidal by design.

Menard, Orville D. *The army and the Fifth Republic.* Lincoln: University of Nebraska, 1967. 265 pp.
An examination of the politicization of the *paras* in the process of waging low-intensity warfare that led to their taking action against the state during the Algerian war. 218 refs.

Millar, George. *The Bruneval raid: Flashpoint of the radar war.* London: Bodley Head, 1974. 208 pp. 43 refs.

Montagnon, Pierre. *La guerre d'Algérie: Genèse et égrenage d'une tragedie.* Paris: Pygmalion/Gerard Watelet, 1984. 452 pp. 70 refs.

Mrazek, James E. *The glider war*. London: Robert Hale, 1975. 304 pp.
A survey of German, Soviet, British, and American military glider development and operations from the 1930s until the 1970s by a veteran of glider-borne operations in World War II.

Narasimhan, R. "The Chindits campaign in Burma." In *Military history anthology*. Fort Leavenworth: Combat Studies Institute, 1984, pp. 87-108.
An Indian army officer's highly critical appraisal of both Wingate and the long-range penetration operations of 1943 and 1944, with considerable attention paid to air operations and special praise for Cochran and Alison. Written in 1973. 8 refs.

Pawle, Gerald. *The secret war, 1939-1945*. New York: W. Sloane Associates, 1957. 297 pp.
A personal account of involvement in the development of weapons for special operations in Britain in World War II.

Phillips, C.E. Lucas. *The greatest raid of them all: An account of the famous World War II commando raid against ships of the German navy*. Boston: Little Brown, 1960. 270 pp.

Piekalkiewicz, Janusz. *Secret agents, spies and saboteurs: Famous undercover missions of World War II*. New York: William Morrow, 1973. 523 pp.
A well-illustrated, popular, large format volume. Not to be used as a source without cross-referencing.

Pointon, Patrick. *They have their exits*. Boston: Little Brown, 1953. 275 pp.
An account of an RAF low-level pinpoint bombing attack which freed many French resistance members from the grisly Gestapo jail in Amiens in World War II.

Pruitt, James. "Green Beret's first blood: North of the DMZ with RT Viper." *Soldier of Fortune* (12:4) April, 1987, pp. 72-75.

Reilly, David R. *The sixth floor*. London: Frewin, 1969. 224 pp.
An account of an RAF low-level pinpoint bombing attack on the Gestapo headquarters in Copenhagen in the Second World War.

Robertson, Terence. *Dieppe: The shame and the glory*. Boston: Little Brown, 1962. 432 pp.
A popular history of the Dieppe raid. 21 refs.

Rundell, Walter, Jr. *Military money: A fiscal history of the U.S. Army overseas in World War II*. 271 pp.
Includes descriptions of tactics of currency use in American special operations in the Second World War.

Ryan, Cornelius. *The longest day.* New York: Simon and Schuster, 1959. 350 pp.

A popular history of D-Day, heavily based on interviews and archival material, including details of various special units' activities on the 6th of June 1944. 133 refs.

Sassar, Charles W. "Secret SOG mission draws first Tet blood: Prairie fire spies Charlie's march." *Soldier of Fortune* (7:4) April, 1982, pp. 32-35.

Schemmer, Benjamin F. *The raid.* New York: Harper and Row, 1976. 326 pp.

An account of the ill-fated Son Tay raid, with close focus on the exploits of Colonel "Bull" Simons, by the editor of *Armed Forces Journal International*, a major advocate of the expansion of special/covert operations in the Reagan years.

Schofield, Stephen. *"Musketoon"--Commando raid: Glomfjord, 1942.* London: Jonathan Cape, 1964. 156 pp.

Sciutti, W.J. "The first air commando group, August 1943 - May 1944." *American Aviation Historical Society Journal* (13:3) Fall, 1968, pp. 178-185.

Short, Anthony. *The Communist insurrection in Malaya, 1948-1960.* New York: Crane Russak, 1975. 547 pp. 109 refs.

Smith, Robert Ross. *Triumph in the Philippines.* Washington, D.C.: Office of the Chief of Military History, 1963. 756 pp.

The U.S. Army's official campaign history; includes accounts of the "triphibious" raids on the Las Baños and Cabanatuan prisoner of war camps, and the airborne assault on Corregidor fortress.

Sochurek, Howard. "Dien Bien Phu drop: The final French loss in Indochina." *Soldier of Fortune* (11:3) March, 1986, pp. 66-73.

----------. "Some dared; some didn't: British SAS plans to raid Argentina." *Soldier of Fortune* (11:3) March, 1986, pp. 46-51.

Stafford, David. "Secret operations versus secret intelligence in World War II: The British experience." In *Men at war, politics, technology and innovation in the twentieth century*, eds. Timothy Traun and Christine Archer. Chicago: Precedent Publishing, 1982, pp. 119-136.

An account of bureaucratic conflict between the British Secret Intelligence Service and the Special Operations Executive in World War II, with priority formally assigned to the former. By a Canadian historian drawing mainly on published accounts and open British records, inasmuch as SIS and SOE archives are closed. 57 refs.

Stanton, Shelby L. *The rise and fall of an army: U.S. government forces in Vietnam, 1965-1973*. Novato: Presidio Press, 1985. 411 pp.
Extensive detailing on *corps d'élite* and related operations. 336 refs.

"Surface action: Submarine special ops in Vietnam." *Soldier of Fortune* (12:5) May, 1987, pp. 44-47.

The symbolic power of air power in some national resistance movements, 1943-1945. Manhattan, Ks.: Sunflower Press, c. 1985. [Reprint of U.S. Air Force Historical Studies No. 12 (1954).] 117 pp.

Taylor, Blaine. "Before invasions commence." *Military History* (4:2) October, 1987, pp. 42-46, and 48-49.
An interview of retired Navy Captain Phil H. Bucklew, a member of the U.S. Navy's Scouts and Raiders in World War II.

Tigges, Duane. "Unsung warriors." *Soldier of Fortune* (4:10) October, 1979, pp. 60-61.

Trenownden, Ian. *Operations most secret: S.O.E., the Malayan theater*. London: Kimber, 1978. 231 pp.
An account of special operations of the Special Operations Executive in Southeast Asia in World War II.

Vaux, Nick (Maj. Gen.). *Take that hill!* Elmsford, N.Y.: Pergamon-Brassey's, 1987. 261 pp.
A history from a participant's view of the operations of No. 42 Royal Marine Commando in the Falklands War. 261 p.

Waddell, Thomas D. et al., eds. *Air war Vietnam*. New York: Arno Press, 1978. 361 pp.
An overview of four major phases of the Vietnamese air war from a U.S. Air Force perspective. Some details on special aircraft and missions, e.g., AC 130 variants, and the role of air force special operations aircraft in the ill-fated *Mayaguez* rescue mission in insertion, support, and extraction of marines on Koh Tang Island.

Wallace, Robert. *The Italian campaign*. New York: Time-Life Books, 1978. 208 pp.
A popular history including a series of photographs of the German airborne forces' rescue of Mussolini in 1944, with Otto Skorzeny prominently featured.

Warren, Harris G. *Special operations: AAF aid to European resistance movements, 1943-1945*. Manhattan, Ks.: Sunflower

Press, c. 1985. [Reprint of U.S. Air Force Historical Studies No. 12 (1954).] 117 pp.

Webb, Daniel J. "Dieppe raid: A blundering, costly success." In *Military history anthology*. Fort Leavenworth: Combat Studies Institute, 1984, pp. 1-12.

A brief analysis of Dieppe from the perspective of an American serving officer, written in 1973. 21 refs.

Weeks, John. *Assault from the sky: A history of airborne warfare*. New York: G.P. Putnam's, 1978. 192 pp.

Vignettes on various airborne forces are interwoven with descriptions of the assault on Crete; operations in North Africa, the Pacific and northwest Europe in World War II; the Far East in the 1950s; the Middle East and Congo in the 1950s and '60s; and Vietnam; large format, well-illustrated. 27 refs.

Whitehead, William. *Echoes of disaster: Dieppe, 1942*, ed. Terence Macartney-Filgate. New York: St. Martin's, 1979. 191 pp.

One of the more recent of many accounts of the controversial, large-scale raid/reconnaissance force in August 1942 against the strongest Nazi fortifications on the Continent, during the major debate amongst the Russians, Americans, and British over the feasibility of a "Second Front."

Wilcox, W.A. *Chindit column 76*. London: Longmans, Green, 1945. 140 pp.

An account of one of the Long-Range Penetration Group's elements in the Burma campaign.

Wilkinson, J.B. "Sneak attack craft in the Pacific." *U.S. Naval Institute Proceedings* (7:3) March, 1947, pp. 279-287. An overview of German, British, and Japanese underwater special operations in World War II.

Willenson, Kim. "Secret wars." In *The bad war: An oral history of the Vietnam war*. New York: New American Library, 1987, pp. 202-223.

An assemblage of interviews with ex-CIA Director William Colby, Colonel William Corson, General Bruce Palmer, Frank Snepp, Jeff Stein, and Colonel John Waghelstein, dealing with special operations from the early 1950s onward.

Wolf, Marv. "Rough & ready rangers: From mountaineering to mysterious missions." *Soldier of Fortune* (7:4) April, 1982, pp. 48-55.

----------. *The X-craft raid*. New York: Harcourt Brace, 1971. 170 pp.

4

Special Operations in Low-intensity Conflict/ Counter-insurgency

Arbuckle, T. "Rhodesian bush war strategies and tactics: An assessment." *Journal of the Royal United Services Institute* (124:4) December, 1979, pp. 27-32.

Description and critique of Rhodesian *corps d'élite* and auxiliaries tactics versus insurgents. Criticizes military leaders' tendency to overweigh military operations versus political and psychological factors.

Barber, Noel. *The war of the running dogs: The Malaya emergency, 1948-1960*. New York: Weybright and Talley, 1971. 283 pp. 43 refs.

Bodansky, Yossef. "Soviet *spetsnaz* at Greenham." *Jane's Defense Weekly* (5:3) January 25, 1986, p. 83.

An article which suggests that Soviet GRU (military intelligence) agents were present as *agents provocateur* and to gain information regarding the installations during the demonstrations against cruise missile deployment. Suggests the Soviets use mockups of such key installations to train their *spetsnaz*; also includes reference to anti-*spetsnaz* aspects of British defense exercise *Brave Defender*, September 1986.

Charters, Davis A. "Special operations in counterinsurgency: The Farran case, Palestine, 1947." *Journal of the Royal United Service Institution* (124:2) June, 1977, pp. 56-61.

Courrier, Yves. *La guerre d'Algériè en images*. Paris: Fayard, 1972. 179 pp.

----------. *Les temps du leopards: La guerre d'Algériè II*. Paris: Fayard, 1971. 613 pp. 96 refs.

----------. *L'fils de la Toussaint*. Paris: Fayard, 1968. 469 pp. 23 refs.

A trilogy on the French ordeal in Algeria.

Dean, David, ed. *Low-intensity conflict and modern technology*. Maxwell Air Force Base: Air University, 1986. 381 pp.

Dobson, Christopher, and Ronald Payne. *The Carlos complex.* London: Hodder Stoughton, 1977. 254 pp.
A description of the Entebbe raid's origins and the tracing of links to a major terrorist elements.

Emerson, Steven. *Secret warriors: Inside the covert military operations of the Reagan era.* New York: G.P. Putnam's, 1986. 256 pp.
A survey of the array of special units and activities that burgeoned in the aftermath of the failure of the Iran hostage rescue raid in 1980, on through the covert operations renaissance in the Reagan era, based mainly on interviews.

Farran, Roy A. *Operation Tombola.* London: Collins, 1960. 256 pp.

Friedman, Richard S. et al. *Advanced technology warfare.* New York: Harmony Books, 1985. 208 pp.
A well-illustrated overview of techniques and equipment is included on pp. 194-205.

Gagan, Pat. "The OV-1D penetration mission." *Military Intelligence* (8:2) April-June, 1982, pp. 32-34.
A rare glimpse at air special operations in low-intensity conflict/covert operations.

Geraghty, Tony. "A new breed of heroes." *Journal of the Royal United Service Institution* (127:1) March, 1982, pp. 69-70.
An overview of operational difficulties and bureaucratic problems against the backdrop of excruciating nervous strain of explosive ordnance demolition work in Ulster in the late 1970s. Perhaps the most rarified of special operations and tests of raw courage among modern *corps d'élite.*

Greenberg, Lawrence M. *The Hukbalahap insurgency: A case study of a successful anti-insurgency operations in the Philippines, 1946-1955.* Washington, D.C.: U.S. Army Center for Military History, Analysis Branch, 1986. 154 pp.

Hansen, Lynn M. *Soviet naval spetsnaz: Operations on the northern flank: Implications for the defense of Western Europe.* College Station, Texas: Texas A&M University, Center for Strategic Technology, 1984. 41 pp.
An analysis of Soviet submarine activities and apparent use of exotic underwater remote systems in the early 1980s, based on Swedish, Russian, and English language sources, by a seasoned American arms control negotiator. 83 refs,

Hayden, H.T. (Lt. Col.). "Antiguerrilla warfare--Soviet style." *U.S. Naval Institute Proceedings* (111:11) November, 1985, pp. 118-120.

Includes data on the role of *spetsnaz* in counterinsurgency operations.

Iran Task Force. "Iran: Consequence of the abortive attempt to rescue the American hostages." *Conflict* (3:1) 1981, pp. 55-78.

James, Harold D., and D. Sheil-Small. *The undeclared war: the story of Indonesian confrontation, 1962-1966*. London: Leo Cooper, 1971. 201 pp.

A plethora of special units in a complex fugue of special operations are encompassed in this account of the pseudo-war which wound down after the toppling of Sukarno. 127 refs.

Jeapes, Tony. *Special air service operation Oman*. London: William Kimber, 1980. 247 pp.

An account of SAS operations in the Dhofar War, 1971-76 by a senior SAS officer. "Lessons" are listed, pp. 228-229. No refs.

Kleite, Immanuel J. "U.S. assistance to Venezuela and Chile in combatting insurgency, 1963-1964--Two cases." *Conflict* (3:4) 1982, pp. 227-244.

United States army unilateral and coalition operations in the 1965 Dominican Republic intervention. Washington, D.C.: U.S. Army Center for Military History, Analysis Branch, 1987. 115 pp.

Includes data on the role of the 82nd airborne division in the Dominican Republic intervention, including training, control, and fire discipline. 210 refs.

Lefever, Ernest W. *Crisis in the Congo*. Washington: Brookings Institution, 1965. 215 pp.

An overview of the complex of mercenaries, *corps d'élite*, and United Nations forces in the events devolving from Katangan attempts at separatism in the early 1960s. 20 refs.

Majdalany, Fred. *State of emergency: The full story of Mau-mau*. Boston: Houghton-Mifflin, 1963. 239 pp.

Describes formation of "special forces" of former insurgents used to track and penetrate other groups.

Manor, Leroy J. (Brig. Gen.). *Report on the Son Tay prisoner of war rescue operation*. Parts I & II. Washington, D.C.: Office of the Joint Chiefs of Staff, Joint Contingency Task Group, 1971.

Morelli, Donald R., and Michael M. Ferguson. "Low-intensity conflict." *International Defense Review* (17:9) 1984, pp. 1219-1220.

"Iran rescue mission: Group analyzes reasons for failure." *Aviation Week and Space Technology*, September 15, 1981, pp. 61-63.

Joint low-intensity conflict project final report: Analytical review of low-intensity conflict. Vol. 1. Fort Monroe, Virginia: U.S. Army Training and Doctrine Command. 325 pp.

Olsen, Jack. *Aphrodite: Desperate mission.* New York: Putnam, 1970. 328 pp.

Patrick, Derrick. *Fetch Felix: The fight against the Ulster bombers, 1976-1977.* London: Hamish Hamilton, 1981.

Patterson, Charles J., and G. Lee Tippin. *The heroes who fell from grace: The true story of Operation Lazarus, the attempt to free American POWS from Laos in 1982.* Canton, Ohio: Daring Books, 1985. 319 pp.

Schemmer, Benjamin. *The raid.* New York: Harper and Row, 1976. 326 pp.
An account of the attempt of U.S. special forces elements under the leadership of Colonel "Bull" Simon to free U.S. prisoners of war at Son Tay, North Vietnam, November 21, 1970. 47 refs.

Sergeev, S.S. "Special operations against socialist forces." In *Totalnyii sphionazh.* Moscow: Voenizdat, 1984, pp. 15-45.
A Soviet condemnation of Western *corps d'élite* during the intensification of the Cold War in the first Reagan administration.

Short, Anthony. *The Communist insurrection in Malaya, 1948-1960.* New York: Crane Russak, 1975. 547 pp. 116 refs.

Stevenson,, William, and Uri Dan. *90 minutes at Entebbe.* New York: Bantam Books, 1976. 216 pp.

Tremayne, Penelope. "End of ten years' war." *Journal of the Royal United Services Institute* (122:1) March, 1977, pp. 44-48.
A review of British-controlled special operations in the Dhofar region of Oman in the late 1960s and early 1970s.

Valeriano, Napoleon D. (Col.), and T.R. Charles (Lt. Col.). *Counter-guerrilla operations: The Philippine experience.* New York: Praeger, 1966. 276 pp.
A history and analysis, based on experience and reflection, of American-supported Philippine counterinsurgency operations in

the late 1940s and 50s, including the role of special units, e.g., Force X. 52 refs.

Wagoner, Fred E. *Dragon rouge: The rescue of hostages in the Congo*. Washington, D.C.: National Defense University, 1980. 219 pp.

Wagoner's account of the rescue by Belgian paratroops of European and American hostages at Stanleyville in the Congo in late November 1964 reveals the complex political, diplomatic, and propaganda dimensions of a special operation, leading to the equally controversial cancellation of phase II of the rescue, thus abandoning hundreds of other hostages to torture and murder in outlying areas. 307 refs.

Wilson, R. Dare (Maj. Gen.). *Cordon and search: With the 6th airborne division in Palestine*. Nashville: Battery Press, 1984. 276 pp.

An account of airborne forces in the security role that they so often played since World War II, in this instance, in Palestine 1946-48, on the eve of the creation of the state of Israel. Extensive appendices present operational data. No refs.

Winkates, James E. "Hostage rescue in a hostile environment: Lessons learned from the Son Tay, Mayaguez and Entebbe missions." In *Political terrorism and business: The threat and response*, eds. Yonah Alexander and Robert A. Kilmore. New York: Praeger Special Studies, 1979, pp. 212-244.

5

Counterterror Operations

Ben Porat, Yesmayamu, Etian Haber, and Zeen Schiff. *Entebbe rescue*, trans. Louis Williams. New York: Delacorte, 1977. 347 pp.

A detailed account of the Israeli long-range, low-altitude attack that freed Israeli hostages in the most complex counterterror special operation to date. No refs.

Buffenbarger, Kurt. "Operation 'Slyfox': Special forces' Jordanian secret." *Soldier of Fortune* (10:4) May, 1985, pp. 32-35.

Burgess, William H., III. "Special operations and the challenge of transnational terrorism." *Military Intelligence* (12:2) April-June, 1986, pp. 8-15, and 48.

Campbell, Doug. "SAS dares & wins. When the talking stopped and the shooting started, the SAS moved in." *Soldier of Fortune* (5:9) September, 1980, pp. 26-29.

Dobson, Christopher, and Ronald Payne. *Counterattack: The West's battle against the terrorist*. New York: Facts on File, c. 1982. 198 pp.

A survey of major antiterrorist forces, operations, and the efforts taken in commercial circles against terrorism. 45 refs.

Fanshaw, Paul. "Target Kolwezi: French Foreign Legion drops into Zaire." *Soldier of Fortune* (8:12) December, 1983, pp. 48-55.

Hoffman, Bruce. "Commando warfare and small raiding parties as part of counterterrorist military policy." *Conflict* (7:1) 1987, pp. 15-45.

Koch, Peter, and Kai Hermann. *Assault at Mogadishu*. London: Corgi, 1977. 165 pp. No refs.

Moessmer, W.C. "Congo's Operation Dragon Rouge: Up in the air over Stanleyville." *Soldier of Fortune* (7:6) June, 1982, pp. 28-33.

Moss, Robert. "Moscow backs terror." *Daily Telegraph* (U.K.) July 16, 1979, p. 5.

A description of culling of foreign trainees for special training in KGB and GRU terrorism schools.

Schoch, Bruce P. "Four rules for a successful rescue." *Army* (31:2) February, 1981, pp. 22-25.

Rules for attempting to free hostages derived from analysis of cases in the 1970s: liberate safely as possible, without giving in to unacceptable demands of captors; keep force size and weapons at the minimum needed for the mission; use forces experienced in working and planning under pressure, and moving fast; gather and use information quickly, i.e., include a strong intelligence component; and encourage sensitivity among forces.

Shortt, Jim. "7 years for 11 minutes." *Soldier of Fortune* (12:12) December, 1987, pp. 32-39.

A detailed description of SAS training and the carrying out of the Prince's Gate operation.

Strasser, Steven, Elaine Sciolino, and Lin Widman. "How Dozier was freed." *Newsweek* (99:6) February 8, 1982, pp. 40-44, and 46.

An account of the first successful freeing of a kidnapped American general from the Red Brigades by Italian security forces after 12 years of activity, including details on Italian special police and *carabinieri* security units.

Thompson, Leroy. *The rescuers: The world's top antiterrorist units*. Boulder: Paladin Press, 1986. 241 pp.

An overview of antiterrorist operations, units, and equipment, with an evaluation of pre-eminent *corps d'élite*, an appraisal of weapons and technical equipment, and lists of units and of terrorist groups. No refs.

Tinnin, David D., with Dag Christensen. *The hit team*. Boston: Little Brown, 1976. 240 pp.

A journalistic account of purported covert Israeli assassinations of "Black September" members in retaliation for killing Israeli athletes at the 1972 Olympics.

6

Biography/Autobiography

Alison, John R. "The most unforgettable character I've ever met: Philip Cochran." *Reader's Digest* (73:438) October, 1948, pp. 130-132, 134, 136, and 138.

A short biography of Cochran who, along with Alison, organized and led the first American "air commando" unit in Burma in World War II; some operational details.

Altieri, James. *The spearheaders: A personal history of Darby's Rangers*. Washington, D.C.: Zenger Publishing, 1979. 325 pp.

An autobiographical history of the Rangers in World War II by a veteran Ranger company commander.

Aspinall-Oglander, Cecil. *Roger Keyes: Being the biography of Admiral of the Fleet Lord Keyes of Zeebrugge and Dow, G.C.B., K.C.V.O., C.M.G., D.S.O.* London: The Hogarth Press, 1951. 478 pp.

Keyes headed Combined Operations from July 1940 to October 1941.

Bank, Aaron. *From O.S.S. to Green Berets*. Novato, Calif.: Presidio Press, 1986. 216 pp.

A memoir of a senior U.S. Army Special Forces officer, from the Second World War through Vietnam, with emphasis on psychological warfare in the early phase of reformation.

Beeson, Colin. *The glider pilot war at home and overseas*. Manhattan, Kansas: Sunflower Press, 1978. 263 pp.

Letters by the author in training and after operation. No refs.

Biggs, Bradley. *Gavin*. Hamden, Conn.: Archon Books, 1980. 182 pp.

A brief biography of "Jumpin' Jim" Gavin, commander of the 82nd airborne division in World War II, and a senior member of the post-war "airborne club." 282 refs.

Blair, Clay. *Ridgeway's paratroopers: The American airborne in World War II*. New York: Dial Press, 1985. 588 pp.

An exhaustively documented and admiring biography of Matthew B. Ridgeway. He headed the 101st airborne division in the European theater in World War II, assumed command of the 8th

Army in Korea in its darkest hours, and went on to be army chief-of-staff and NATO commander. This account is, however, little inclined to criticality of its subjects. An excellent bibliography. 1,357 references.

Blankfort, Michael. *The big Yankee: The life of Carlson of the raiders*. Norton: Little Brown, 1947. 380 pp.

A hagiography of Evans Carlson, who trained the U.S. Marine Corps 2nd Raider Battalion in World War II, using principles derived from observing the Chinese Communist 8th Route Army in the late 1930s. Marine Corps senior officers were not enthusiastic but President Roosevelt, and his son James, a Marine reserve office who became executive officer of the 2nd, wanted a U.S. version of British commandos. The film "Gung Ho" glamorized the 2nd Raiders and provided a pulpit for Carlson's populist views. He was later wounded and the Raiders were folded into a reconstituted 4th Marine regiment. He became a brigadier general, died soon after the war, and was a target for criticism in U.S. debates over China policy.

Borghese, J. Valerio. *Sea devils*, trans. James Cleugh. Chicago: Henry Regnery, 1954. 261 pp.

An account of wartime services with the 10th M.A.S. by an Italian officer, including details of development of equipment and operations and descriptions of plans for attacks on Freetown and New York and a major assault on Gibraltar, all cancelled when Italy left the war.

Bradley, Omar. *A soldier's story*. New York: Henry Holt, 1951. 618 pp.

Bradley, as 12th Army group commander, was the senior U.S. ground commander in Northwest Europe, under Montgomery from D-Day to D-Day Plus 90, then under Eisenhower. Holding a view similar to Mark Clark's of airborne forces having too much "down-time," he ordered the overrunning of drop zones to undercut the airborne *raison d'être*.

Caldwell, Bob. "Aaron Bank saga: 'Tell Bank to get Hitler!'" *Soldier of Fortune* (7:6) June, 1982, pp. 24-27.

Carter, Ross S. *Those devils in baggy pants*. New York: Appleton-Century-Crofts, 1955. 192 pp.

A classic personal account of combat experiences by a member of the 504th parachute infantry regiment of the 82nd U.S. airborne infantry division.

Casper, Bernard M. *With the Jewish brigade*. London: Edward Goldston, 1947. 128 pp.

The wartime memoir of a chaplain with detailing of the formation of the unit that attracted the "cream" of Jewish Palestinian youth, and served as a cadre for the development of the Israeli armed forces after World War II.

Chennault, Claire L. *Way of a fighter: The memoirs of Claire Lee Chennault*. New York: G.P. Putnam's, 1949. 375 pp.

A fighter advocate in the bomber-dominated U.S. Army Corps in the 1930s, Chennault was medically retired, and after serving as a mercenary pilot for the Nationalist Chinese, he helped to raise and commanded the American Volunteer Group--the "Flying Tigers"--and after their disbandment in 1942, he became commander of the 14th Air Force in the China-Burma-India theater. Much interesting data, but must be compared with other principal sources on the China-Burma-India campaign.

Clarke, A.F.N. *Contact*. London: Pan Books, 1983. 162 pp.

A controversial memoir, by a British parachute regiment veteran, of 1973 and 1976 tours in Ulster.

Clarke, Dudley. *Seven assignments*. London: Jonathan Cape, 1948. 262 pp.

Clarke's narrative of his wide-ranging activities from Africa to Norway in the opening months of World War II concludes with an account of this role in the founding and subsequent operations of the commandos. As a staff officer under then Chief of the Imperial General Staff Sir John Dill, Clarke did the staff work in response to Prime Minister Winston Churchill's call for aggressive initiatives after the Dunkirk evacuation in 1940. Clarke was thereafter involved in organizing the first raids under M.O. 9, by No. 11 Commando, drawn mainly from the independent companies raised from territorial divisions for Norway. Included are references to film actor David Niven who returned to active service in World War II, having been a regular officer in the early 1930s.

Colby, William, and Peter Forbath. *Honorable men: My life in the C.I.A.* New York: Simon and Schuster, 1978. 493. pp.

A memoir by an ex-director of the CIA who served in the O.S.S. in World War II in paramilitary/special operations. Perhaps the strongest and most experienced director since Allen Dulles, he strongly favored maintaining covert capabilities.

Cowles, Virginia S. *The phantom major: The story of David Stirling and his desert command*. New York: Harper and Brothers, 1958. 320 pp.

A biography of Stirling and a history of SAS operations in North Africa, written mainly from interviews and correspondence with veterans of the campaign.

Darby, William Orlando, with William H. Baumer. *Darby's rangers: We led the way*. San Rafael: Presidio Press, 1980. 198 pp.

A co-founder of the U.S. Army's World War II Ranger force, Darby led the largest contingent in combat, which suffered the highest loss rate of an American *corps d'élite* in the 20th

Century. Darby was killed in action soon afterward while serving with the 10th Mountain Division.

Durnford-Slater, John. *Commando*. London: William Kimber, 1953. 222 pp.
An autobiographical account by a British brigadier, a regular gunner, of his service with No. 3 Commando in World War II, including chapters on training and operations, including Guernsey, the Löfotens, Väagso, Dieppe, North Africa, Sicily, Italy, Overlord, and Northwest Europe.

Edell, Ed. *A special breed of man*. Guilderland, NY: Ranger Associates, 1984. 241 pp.
A biography in popular style of a special forces NCO, focusing on the 2nd Indochina War. No refs.

Farran, Roy. *Winged dagger: Adventures on special service*. London: Collins, 1948. 384 pp.
A memoir by an SAS veteran of his service in North Africa, France, and in post-war British imperial/Commonwealth operations.

Farrar-Hockley, A.H. *Student*. New York: Ballantine Books, 1973. 160 pp.
The Ballantine World War II series biography of Kurt Student, the main shaper of German airborne forces. 14 refs.

Foley, Charles. *Commando extraordinary*. New York: Ballantine Books, 1955.
A biography of Otto Skorzeny.

Fraser, Ian. *Frogman V.C.* London: Angus and Robertson, 1957.

Gale, Richard. *Call to arms: Autobiography*. London: Hutchinson, 1968. 280 pp.
Includes two chapters on Gale's role as commander of the 1st British airborne brigade during its formation, of the British 6th airborne division in operation *Overlord*, and of the XVIIIth Airborne Corps; also of pertinence is his defensive rationale for *corps d'élite* on p. 117. No refs.

Gardner, Arthur D. "The cream of the corps: Marine raider's Guadalcanal diary." *Soldier of Fortune* (6:1) January, 1981, pp. 58-67.

Gavin, James M. *On to Berlin: Battles of an airborne commander, 1943-1946*. New York: Viking Press, 1987. 336 pp.
A war memoir by a young cavalry captain, one of many who built the American airborne program from scratch during the major pre-war build-up, and rose to command the 82nd airborne division in Europe. Gavin later became chief of army research

and development, resigned in protest over the Eisenhower cuts, and became ambassador to France in the Kennedy administration. While head of Arthur D. Little in the late 1960s, his proposal for enclaves and a strategy of defensive pacification led to his being mentioned as a presidential candidate.

Gehlen, Reinhard. *The service: The memoirs of Reinhard Gehlen*, trans. David Irving. New York: World Publishing, 1972. 386 pp.

A work much criticized for its vagueness and defensiveness in the wake of major intelligence scandals in the German Federal Republic's *Bundesnachrichtdienst* [Intelligence Service], which Gehlen formed after serving as head of the German army's Foreign Armies East section in World War II, and in the United States' intelligence nexus after the war. Includes references to special operations plans, units, and operations in the Second World War.

"General John K. Singlaub." *Soldier of Fortune* (4:1) January, 1979, pp. 30-31.

Glubb, John Bagot. *A soldier with the Arabs*. London: Hodder and Stoughton, 1957. 458 pp.

One of many works by "Glubb Pasha." An account of Glubb's role in raising what was viewed as the highest quality Arab fighting force in the Middle East until at least the late 1960s.

Gordon, John, IV. "Among the Best: The Philippine Scouts." *Military Review* (57:9) September, 1987, pp. 69-79.

Grivas, George. *The memoirs of General Grivas*, ed. Charles Foley. New York: Praeger, 1965. 226 pp.

The memoirs of the principal Greek nationalist underground leader on Cyprus during the 1950s, including details on his World War II special operations experience, the Greek Civil War, and his role as EOKA leader.

Halpern, Paul G., ed. *The Keyes papers: Selections from the private and official correspondence of Admiral of the Fleet Baron Keyes of Zeebrugge*. Vol. III, 1939-1949. London: George Allen and Unwin, 1981. 398 pp.

The first two-thirds of Part II deals with Keyes' service as Director of Combined Operations, July 1940-October 1941, and his bureaucratic battles, operational frustrations, and falling out with Churchill. His son Geoffrey was killed in 1941 during the abortive commando raid on Rommel depicted in the opening sequence of the film "Rommel, the Desert Fox."

Hamilton-Hill, Donald. *S.O.E. assignment*. London: William Kimber, 1973. 186 pp.

A memoir of service with special operations executive auxiliary units, which operated as cadres for resistance forces in World War II in France.

Henderson, Ian, with Philip Goodhart. *Man hunt in Kenya.* Garden City: Doubleday, 1958. 240 pp.

The account of a Kenyan-born settler of British ancestry whose fluency in Kikuyu aided him in tracking down gangs during the "Mau-mau" emergency in the early 1950s, and in capturing Dedan Kimathi. "Pseudo-gangs" suggestive of the X-Force in the Philippines are described.

Hough, Richard. *Mountbatten: Hero of our time.* London: Weidenfeld and Nicholson, 1980. 290 pp.

A brief biography of Admiral of the Fleet Lord Louis Mountbatten who headed Combined Operations, 1941-43. 174 refs.

Hutton, Bernard J. *Frogman spy: The incredible case of Commander Crabb.* New York: McDowell-Obolensky, 1958. 180 pp.

A racy account of the much-speculated fate of Commander Lionel Crabb who failed to return from scrutinizing a visiting Soviet warship, based on a purported dossier of interrogation of Crabb, and his transition to Soviet naval officer status.

Infield, Glenn. *Skorzeny: Hitler's commando.* New York: St. Martin's Press, c. 1981. 266 pp.

A popular biography of the *SS* officer whose rescue of Mussolini, kidnapping of Admiral Horthy, and activities in the Battle of the Bulge were the most visible Nazi special operations in World War II. The first third deals with wartime operations, and the remainder with complex postwar purported neo-Nazi activities and links. 217 refs.

Lansdale, Edward Gerry. *In the midst of war: An American mission to Southeast Asia.* New York: Harper and Row, 1972. 386 pp.

A memoir of the pre-eminent U.S. Air Force/CIA theoretician and practitioner of counterinsurgency, covering his experiences in the Philippines in the late 1940s and early 1950s, and in Vietnam through 1955.

Lawson, Ted. *Thirty seconds over Tokyo*, ed. Robert Considine. New York: Random House, 1943. 221 pp.

Autobiographical account of the "Doolittle Raid," a special operation in essence, aimed at enhancing civilian morale and demonstrating aggressiveness when conventional operations were going badly. Navy and Army Air Corps elements combined to launch an attack on Japanese cities in a high risk operation in April 1942. Most crews survived, and Lawson's account was the basis of a major motion picture starring Van Johnson and Spencer Tracy in 1944.

Leuliette, Pierre. *St. Michael and the dragon: Memoirs of a paratrooper*, trans. John Edmonds. Boston: Houghton Mifflin, 1964. 443 pp.

A memoir of service in Algeria, 1954-57, with details on "tribal" aspects of the *paras* and on the use of torture; the latter caused the book to be banned in France at the time.

Lloyd-Owen, David. *Providence their guide*. London: Harrap, 1980. 238 pp.

A memoir by a veteran of Long Range Desert Group operations in North Africa in World War II.

MacCloskey, Monro. *Secret air missions*. New York: Richard Rosen, 1966. 169 pp.

A collection of letters and a personal narrative by the commander of a squadron of the 15th U.S. Air Force detailed to support clandestine operations in the Mediterranean in World War II.

Marrinan, Patrick. *Colonel Paddy*. Dungannon: The Ulster Press, 1968. 207 pp.

A biography of Paddy Mayne of the SAS, who destroyed far more enemy aircraft on the ground in North Africa in World War II than most combat pilots did in the air.

Martin, Harold H. "Paratrooper in the Pentagon." *Saturday Evening Post* (227) August 28, 1954, pp. 23-24.

A profile of Gavin during his service as chief of U.S. Army research and development.

Masters, John. *The road past Mandalay*. New York: Harper and Row, 1961. 341 pp.

An account of Masters' experiences in the Burma campaign where he served as a staff officer and then as a brigade commander of the Chindits as they were ground down in combat, and under the political pressure from "Vinegar Joe" Stilwell, who, like a number of other major commanders, held lightly equipped and supported *corps d'élite* in action well beyond their capacity, resulting in inordinately heavy casualities.

Massu, Jacques. *Le torrent et la digue*. Evreux: Plon, 1972. 405 pp.

----------. *La vraie bataille d'Alger*. Evreux: Plon, 1971. 392 pp.

Memoirs of the pre-eminent airborne commander who played a key role in the near-coup that brought de Gaulle to power and who, with Raoul Salan, orchestrated the controversial low-intensity warfare campaign. Deals with the question of torture which led to popular demand for French withdrawal. (Some photographs in *La vraie. . . .* are not for the squeamish.)

Mead, Peter. "Orde Wingate and the official historian." *Journal of Contemporary History* (14:1) January, 1979, pp. 55-82.
A *defensa* of Wingate's reputation against strong judgments levied in the British offical history of the Burmese campaign. For a reaction by Brigadier Shelford Bidwell, and rebuttal by Mead, see *JCH*, 15:2 and 15:3.

Morris, Jim. "Singlaub's war: With the OSS in the CBI theater." *Soldier of Fortune* (12:2) February, 1987, pp. 68-71.

----------. "Tales from Uncle Jack: The saga of General Singlaub." *Soldier of Fortune* (11:10) October, 1986, pp. 60-64.
Singlaub, a veteran of special operations, emerged into public view when commander of U.S. forces in Korea in the Carter administration; and later as a witness in the Iran-Contra hearings.

Mosley, Leonard. *Gideon goes to war*. New York: Charles Scribner's Sons, 1955. 256 pp.
A popular biography of Orde Wingate, who helped Palestinian Jewish settlers from the "Special Night Squads" in the late 1930s which later evolved into the *Palmach*, the elite cadre of the Israeli army. Wingate was then involved in the toppling of the Italian East African Empire in 1941, and after gaining Churchill's mentorship, he developed the long range penetration concept carried out by the Chindits and, later, by Merrill's Marauders in Burma, 1943-44. He was killed in a plane crash in 1944.

Murray, Simon. *Legionnaire*. Paris: Pygmalion/Gerard Watelt, 1984. 320 pp.
An autobiographic account by a British foreign legionnaire of his service, 1960-65.

Nationale front des demokratischen deutschland. *Brown book: War and Nazi criminals in West Germany*. Leipzig: Verlag Zeit im Bild, c. 1956. 402 pp.
Includes condemnatory material on Skorzeny as head of "*der Spinne*."

Netanyahu, Benjamin, and Iddo Netanyahu, eds. *Self-portrait of a hero: The letters of Jonathan Netanyahu (1963-1976)*. New York: Random House, 1980. 304 pp.
Rare insight into the complex thoughts and concerns of a veteran Israeli paratroop leader killed while commanding the rescue of hostages at Entebbe through his letters to friends, family, and fiancée.

Niven, David. *The moon's a balloon*. New York: G.P. Putnam, 1972. 348 pp.
The memoir of a famous British film star of the Hollywood Golden Age, who was regular officer (Highland Light Infantry) in

the early 1930s and returned to active service in World War II, including three years with Phantom after commando training, liaison between M.O.9 and commando units, and detached to film-making service.

Owen, David Lloyd. *The desert my dwelling place.* London: Cassell, 1957. 271 pp.

----------. *Providence their guide. A personal account of the Long Range Desert Group 1940-1945.* London: Harrap, 1980. 238 pp.

Pugh, Marshall. *Frogman: Commander Crabb's story.* New York: Scribner's, 1956. 208 pp.

A popular, spare, and speculative account of the pre-eminent British World War II underwater warfare expert, who vanished during the visit of the Soviet cruiser *Ordzhonikidze* to Britain during the coronation of Elizabeth II, concluding that he died in an attempt to reconnoiter the ship.

Rhodes, James Richard. *Chindit.* London: John Murray, 1980. 214 pp.

Memoirs of a Chindit officer who served under John Masters in Burma in 1944.

Ridgeway, Matthew B. *Soldier: The memoirs of Matthew B. Ridgeway.* New York: Harper and Brothers, 1956. 371 pp.

The autobiography of the most distinguished member of the "airborne club." Ridgeway, perhaps the most able American field commander in the twentieth century, commanded the 82nd airborne division in World War II, the 8th Army in the Korean War, and went on to be NATO supreme commander and U.S. Army chief of staff.

Rossetto, L. *Major General Orde Charles Wingate and the development of long range reconaissance.* Manhattan, Ks.: Sunflower Press, 1982. 492 pp.

Sadler, Barry, and Tom Mahoney. *I'm a lucky one.* New York: Macmillan, 1967. 191 pp.

Memoir of the author of the popular song of the 1960s "The Ballad of the Green Berets," who served as a special forces noncommissioned officer in the 2nd Indochina War.

Schemmer, Ben. "Requiem for a warrior." *Soldier of Fortune* (4:11) November, 1979, pp. 34-37.

Schuon, Karl. *U.S. Marine Corps biographical dictionary.* New York: Franklin Watts, 1963. 278 pp.

Includes biographies of Merritt Edson, commander of the 1st Raider Battalion, and Evans Carlson, of the 2nd.

Servan-Schreiber, Jean-Jacques. *Lieutenant in Algeria*, trans. Ronald Matthews. New York: Alfred A. Knopf, 1957. 231 pp.

Servan-Schreiber, a political journalist and reservist, describes his service in Galland's Nomad Commandos/Black Commandos, an attempt to link with the Arab population versus the general policy of laying on heavy military equipment, manpower, firepower, and related posturing. His critical language regarding the high command and references to brutality and racism added to the furor in France that led to de Gaulle coming to power.

Shavit, Maty. *On the wings of eagles: The story of Arik Sharon, commander of the Israel paratroopers*. Tel-Aviv: Olive Books, 1972. 160 pp.

A biography of one of the luminaries of the central *corps d'élite* of the Israeli army.

Skorzeny, Otto. *Secret missions: war memories of the most dangerous man in Europe*, trans. Jacques LeClerc. New York: E.P. Dutton, 1950. 256 pp.

Slim, William. *Defeat into victory*. London: Cassell, 1956. 576 pp.

The architect of the reverse of arms in Burma as commander of the British XIVth Army, Slim has been to date the most articulate and quoted critic of the proliferation of "mobs for jobs" in the Second World War; see pp. 546-549.

Sosabowski, Stanislaw. *Parachute general*. London: Kimber, 1961. 159 pp.

The memoir of the commander of the Polish parachute brigade committed at Arnhem in a vain attempt to extricate the doomed British 1st airborne division in September 1944.

Spiller, Roger, Joseph G. Dawson III, and T. Harry Williams, eds. *Encyclopedia of American military biography*. 3 vols. Westport, Conn.: Greenwood, 1984. 1,368 pp.

Includes entries on Evans Carlson and Lucian Truscott, as well as several air and airborne commanders.

Taylor, Blaine. "Skorzeny: Germany's legendary warrior." *Soldier of Fortune* (10:6) June, 1985, pp. 66-69.

----------. "More on Skorzeny: Germany's classic commando." *Soldier of Fortune* (10:7) July, 1985, 70ff.

Thompson, Leroy. "British Bedouin: Arab legion leader learned the way of the desert." *Soldier of Fortune* (8:5) May, 1983, pp. 28-31.

Trinquier, Roger. *Le temps perdu* [The lost times]. Paris: Albin Michel, 1978. 442 pp.

A memoir of Colonel Trinquier's experiences in colonial and counter insurgency, including Indochina in World War II and the First Indo-china War, Algeria and Katanga; includes an epilogue on lessons learned. Trinquier's theoretical works include *La guerre moderne and Guerre--subversion--revolution.*

Truby, J. David. "Vietnam visionary: Highest ranking American to die in 'Nam . . . J.P. Vann did it his way." *Soldier of Fortune* (5:1) January, 1980, pp. 68-79.

Truscott, Lucian K., Jr. *Command missions: A personal story.* New York: E.P. Dutton, 1954. 570 pp.

A memoir of World War II by one of the most dynamic senior combat commanders in the European theater in World War II, who played a major role in the creation of the Rangers.

Urquhart, R.E. (Maj. Gen.). *Arnhem.* Derby, Conn.: Monarch Books, 1960. 174 pp.

A staunch *defensa* by the commander of the British 1st airborne division of his and the unit's role in the ill-fated attempt to flank German defenses in Holland in the fall of 1944 in World War II, in which the 1st suffered over eighty percent losses.

Vandegrift, A.A. (Gen.), and Robert D. Asprey. *Once a marine.* New York: W.W. Norton, 1964. 448 pp.

Includes Vandegrift's hostile views toward marine special units and raiders, which, after he became Marine Corps commandant, were disbanded and merged into the reconstituted 4th regiment in 1944, to replace the unit captured by the Japanese when the Philippines fell in 1942.

Volckman, R.W. *We remained: Three years behind the enemy lines in the Philippines.* New York: W.W. Norton, 1954. 244 pp.

A memoir of Volckman's role as head of U.S. armed forces in the Philippines, the network of intelligence, guerrilla, and special forces that operated on Luzon during the Japanese occupation. The last chapter is a treatise on the relationship of psychological warfare and guerrilla/special operations. No refs.

Walker, Martin. *The national front.* London: Fontana, 1977. 224 pp.

Describes post-World War II activities of David Stirling, founder of the SAS, including a private firm offering *corps d'élites* for hire, and involvement in British far right-wing politics.

Weingartner, James J. "Sepp Dietrich, Heinrich Himmler and the Leibstandarte SS Adolf Hitler." *Journal of Central European History* (1:3) September, 1968, pp. 264-284.

A discussion of the links between two of the principals in the *SS*/Gestapo nexus and Hitler's bodyguard unit.

Whiting, Charles. *Skorzeny*. New York: Ballantine Books, 1972. 160 pp. 10 refs.

Young, Peter. *Bedouin command*. London: William Kimber, 1956. 203 pp.

Memoirs of a British officer seconded to the Arab Legion, with extensive detailing on customs, uniforms, and operations up through the 1948 war with Israel.

Ziegler, Philip. *Mountbatten*. New York: Alfred Knopf, 1985. 784 pp.

The definitive biography of Lord Mountbatten. Chapters 11 - 23 cover his various dealings with special units and warfare while heading combined operations, and then as the commander-in-chief of Southeast Asia Command in World War II. 2,738 refs.

7

Bibliographies

Arkin, William M. *Research guide to current military and strategic affairs*. Washington, D.C.: Institute for Policy Studies, 1981. 232 pp.

A broad-ranging guide to research materials, including lists of newsletters, serials, monographs, U.S. military manuals, and area studies.

Becker, Fritz. *Bibliographie: 10 jahre deutscher militräverlag, 1956-66* [Bibliography: 10 years of German military publishers]. Berlin: Deutscher Militräverlag, 1962. 207 pp. 280 items.

Bekker, Cajus. *The Luftwaffe war diaries*, trans. Frank Ziegler. Garden City: Doubleday, 1968. 399 pp.

Includes a basic bibliography of wartime and more recent German sources on various air operations in World War II.

Bibliography, studies in low-intensity conflict, M-771/3, Course 7, Appendix 2. Fort Leavenworth: U.S. Army Command and General Staff College, 1982. pp. LP2-2-1 - LP2-2-7.

A broad-ranging bibliography on guerrillas and insurgency.

Blackstock, Paul W., and Frank L. Schaf, Jr., eds. *Intelligence, espionage, counterespionage and covert operations: A guide to information sources*. Detroit: Gale Research, 1978. Covert operations, Part IV, pp. 189-211. 79 items.

Coletta, Paolo E. *A bibliography of American naval history*. Annapolis: U.S. Naval Institute, 1981. 453 pp. c. 5,000 items.

----------. *An annotated bibliography of U.S. Marine Corps history*. Lanham, Md.: University Press of America, 1986. 423 pp. c. 4,000 items.

Cline, Marjorie W., Carla E. Christiansen, and Judith W. Fontaine, eds. *Scholar's guide to intelligence literature: Bibliography of the Russell J. Bowen collection in the Joseph Mark Lauinger Memorial Library, Georgetown University*. Frederick, Md.: University Publications of America, 1983. 236 pp. c. 6,000 items.

Constantinides, George. *Intelligence and espionage: An analytical bibliography.* Boulder, Co.: Westview, 1985. 559 pp.

Extensive, annotated bibliography with strong evaluative comments and "insider" perspective. Alphabetically ordered. 584 items.

Devore, Ronald M. *Spies and all that: Intelligence agencies and operations: A bibliography.* Los Angeles: California State University, Center for the Study of Armament and Disarmament, 1982. 63 pp.

Not annotated. 556 items.

Dornbusch, Charles Emil. *Austrian military bibliography.* Cornwallisville, N.Y.: Hope Farm Press, 1967. 80 pp.

Ellis, John. *The sharp end: The fighting man in World War II.* New York: Scribners, 1980. 396 pp.

While paying relatively little attention to special units and operations, the bibliography on World War II includes an extensive section on eyewitness accounts. 784 items.

Fall, Bernard. *Street without joy.* Harrisburg: Stackpole, 1961. 408 pp.

Includes a "Military Bibliography of Indochina," dealing with the 1st Indochina War.

Foreign nations: Special studies, 1962-1985. Frederick, Md.: University Publications of America, various dates. Microfilm.

A series of U.S. government-commissioned analyses, including substantial material on low-intensity conflict.

Harris, William R. *Intelligence and national security: A bibliography with selected annotations.* 3 vols. Cambridge, Mass.: Center for International Affairs, 1968. 838 pp.

Higham, Robin. *Official histories: Essays and bibliographies from around the world.* Manhattan: Kansas State University, 1970. 640 pp.

Janke, Peter, and Richard Smith. *Guerrilla and terrorist organizations: A world directory and bibliography.* Brighton: Wheatsheaf Books, 1983. 531 pp.

Kinnell, Susan K., ed. *Military history of the United States: An annotated bibliography.* Santa Barbara, Ca.: ABC-CLIO, 1986. 333 pp. 3,285 items.

Lang, Kurt. *Military institutions and the sociology of war: A review of the literature with annotated bibliography.* Beverly Hills, Ca.: Sage, 1972. 337 pp. 1,325 refs.

Martin, Michel L. *Warriors to managers: The French military establishment since 1945*. Chapel Hill: University of North Carolina, 1981. 424 pp.
Includes an extensive bibliography on French military and defense affairs. 404 items.

Mickolus, Edward, with Peter Flemming. *Terrorism, 1980-1987, a selected bibliography*. Westport, Conn.: Greenwood, 1988. 314 pp.

Morison, Samuel Eliot. *History of United States naval operations in World War II*. Supplement and general index. Boston: Little Brown, 1962. 359 pp.
Includes 64 entries on underwater demolition teams.

New modes of conflict: Urban guerrilla warfare and international terrorism. A bibliography of selected RAND publications. Santa Monica, Ca.: RAND Corporation. 12 pp. 52 items.

Operations Dragon Rouge et Noir (Stan-Paulis, November 1964). Brussells: Service de L'Historique, 1966. 6 pp.
Mainly European-published works on the role of Belgian paratroops in the Congo in 1964. 68 items.

Palmer, Robert, ed. *Bibliography of intelligence literature: A critical and annotated bibliography of open-source literature*, 7th ed. Washington: Defense Intelligence School, 1981. 78 pp. 269 items.

Parrish, Michael. *The Soviet armed forces: Books in English, 1950-1967*. Stanford: Hoover Institution Press, 128 pp.

Peake, Louis A. *The United States in the Vietnam war, 1943-1975: A selected, annotated bibliography*. New York: Garland Publishing, 1986. 406 pp.
Includes items on special operations and *corps d'élite* throughout, and sections on strategy and tactics, counterinsurgency, and psychological-political warfare. 1,447 refs.

Researcher's Guide to Washington. Washington, D.C.: Washington Researchers, 1978. 813 pp.

Rocca, Raymond G., and John J. Dziak, eds. *Bibliography on Soviet intelligence and security services*. Boulder: Westview, 1985. 203 pp. 518 items.

Smith, Myron J. *Air war bibliography, 1939-1945*. 5 vols. Manhattan, Ks.: Sunflower Books, 1978-82. English language sources. 1,394 pp.

Smith, Myron J. *War story guide: An annotated bibliography of military fiction*. Metuchen, N.J.: Scarecrow Press, 1980. 437 pp. 3,917 items

----------. *The secret wars*. Santa Barbara: ABC-Clio. Vol. 1, 1980: *Intelligence, propaganda and psychological warfare, resistance movement and secrets operations, 1939-1945*. 250 pp. Vol. 2, 1981: *Intelligence, propaganda and psychological warfare, covert operations, 1945-1980*. 389 pp.

Vol. 1 includes many references on *corps d'élite* and special operations throughout: eight main categories include "Special Forces Military Units" and "Secret Operations." 2,539 items. Vol. 2 contains no specific headings as in Vol. 1; material distributed under four major headings. 4,713 items.

Terrorism. A bibliography of selected RAND publications. Santa Monica, Ca.: RAND Corporation, 1987. 25 pp. 99 items.

Terrorism: Special studies, 1975-1985. Frederick, Md.: University Publications of America, 1986. Microfilm, 5 reels.

U.S. Naval Institute Proceedings cumulative index, 1874-1977. Annapolis, Md.: U.S. Naval Institute Proceedings, 1982. 377 pp.

Yokota, Marilyn B. "Bibliography on the 1982 Falklands War." *Conflict* (6:2) 1985, pp. 177-190.

Includes references on special operation and units. 258 items.

8

Official Sources

The clandestine nature of special operations makes access especially difficult for researchers, in spite of the roll-back in the 1970s in the date of opening of most sensitive Anglo-American official records from 50 to 30 years after the date of origin. Attempts to retrieve documents from archives or from control of originating agencies under the Freedom of Information Act in the United States can be expensive and frustrating and, often, futile. Nevertheless, American rules are somewhat easier than other nations', where military historians, academicians, and journalists may be under direct state control and patronage. Even in democracies, they may shy away from using unclassified information to seek out patterns or make conjectures regarding information that may be secret and thus stumble afoul the law, or risk unpopularity.

Under pertinent security legislation, there are a number of reasons why an American federal government agency can hold back data. The Freedom of Information Act only allows a researcher to formally ask the controller to determine if a document is appropriately classified. If the agency believes it is, there is no recourse. A route less travelled, but with more potential, is asking for mandatory review under the provisions of Executive Order 12356 which requires classification status of a document to be reviewed. (Only American citizens can request this.)

While this research guide is designed primarily for American researchers and analysts, before considering the basic ranges and collections available in the U.S., a brief consideration of foreign records is worthy of mention. In Britain, there are many documents scattered throughout the various defense categories, e.g., Air, Admiralty and War Office, World War II Premier ranges and Foreign Office records, most of them located at the Public Record Office at Kew, west of London. British security practices are more stringent than in the United States, and there have been reports of recent reclassification of open records. On occasion, titles of sealed files are listed when beyond the limits of the 30-Year Rule, and dates of authorized opening. As in the United States, collections of papers can be found throughout the country in Universities and in private hands. In instances where they are sensitive, they are subject to the pertinent laws and practices of review by various government authorities. The records of the principal secret intelligence and counterintelligence agencies are not accessible. Researchers have sometimes found declassification and speed in handling faster away from main archives.

Some countries, like France, are far more stringent in respect to opening dates, e.g., the death of principals plus seventy-five years. East Germany has a reputation for both special capriciousness and constraints in respect to granting access to archives. There is a general sense of vetting at work in respect to who is allowed access to Soviet records, but with no more of a clear pattern than can be limned in respect to other Russo-Soviet behaviors, before or during *glasnost*.

Records, of course, may be altered, forged, destroyed, or stolen, for reasons ranging from protection of national security to personal whim. Special operations and *corps d'élite* overlap into the domain of intelligence, where anxieties are especially keen, egos especially strong, and political pitfalls especially deep. Hence there is a much greater warping of history, as well as a substantial amount of activity which is never documented, e.g., unvouchered funds and unwritten orders. Some secret documents may be sequestered for generations, e.g., certain blocks of evidence used by the Warren Commission, and Washington's espionage papers, opened in the late 19th Century. A useful guide in this area is the Consortium for the Study of Intelligence's *A Guide to the Declassified Documents Reference System* (1982).

OFFICIAL HISTORIES

Space does not allow full listing and notation of official histories, although some are included in the bibliography. Researchers will want to consult Robin Higham's various guides to British and American military history and official histories. Most campaign and specialized volumes have some description of special units and operations and/or documents citations. For example, one can find extensive references to the American Volunteer Group, Chindits, Air Commandos, Merrill's Marauders, and related personalities in Charles F. Romanus and Riley Sunderland's trilogy on Stilwell, and other perspectives in the British official histories, S. Woodburn Kirby's *The War Against Japan* and Stephen Roskill's *The War at Sea trilogy*, 1939-1945, and in Frank Wesley Craven and James Lea Cate's seven volume *The Army Air Forces in World War II*. Some may not find it too surprising that Soviet official history is somewhat less detailed in regard to special units and operations.

The following document ranges in the National Archives of the United States, therefore, are offered to aid the researcher in the spirit of "ask; but don't expect too much," are, as the bibliography itself, suggestive rather than exhaustive. Records Group Numbers are in the National Archives in Washington proper, unless otherwise indicated:

Navy Record Groups:

38 - COMINCH-CNO Files
80 - SECNAV-CNO (Commander-in-Chief/Secretary of the Navy) Letter Correspondence Files, 1940-1947; Chief of Naval Operations/Vice Chief of Naval Operations Files up to 1954
313 - Records of Naval Operating Forces, 1940-1967 "Flag Files" - at Suitland, Maryland - Includes Operational Plans - done and not done

Army Records:

War Department Records Groups:
165 - General and Special Staffs; includes: psychological warfare material General Staff Operation Plans Division (OPD) G-2 - Assistant Chief of Staff, Intelligence - propaganda files (some in National Archives, D.C.; some at Suitland) G-3 - Organization and training
218 - Records of the Joint Chiefs of Staff prior to 1961
226 - Records of the Office of Strategic Services
319 - Records of the Army Staff - up to 1960; break in 1947 due to National Defense Act creating the Air Force; includes "Cover and Deception" file; 1955 - G-3 became "Dezzops" - DCSOPS - Deputy Chief of Staff for Operations; also includes Office of the Chief of Psychological Warfare
330 - Office of the Secretary of Defense, 1947-1954
331 - SHAEF (Supreme Headquarters Allied Forces Europe); includes 12th and 6th Army Groups and Psychological Warfare Branch
349 - Joint Commands, including U.S. Armed Forces Far East (MacArthur's command); other army records including material related to special operations and corps d'lite are those of the Adjutant General's Office, the Medical Corps, and the Office of the Surgeon General.

The U.S. Marines in Vietnam historical series, published by History and Museums Division, headquarters, United States Marine Corps, Washington, D.C., includes extensive detail on special operations and units and are listed in order of operational sequence.

Whitlow, Robert M. *U.S. Marines in Vietnam: The advisory and combat assistance era, 1954-1964*. Publ. 1977.

Shulimson, J., and C. Johnson. *U.S. Marines in Vietnam: The landing and the build-up, 1965*. Publ. 1978. 261 pp. 456 refs.

Shulimson, Jack. *U.S. Marines in Vietnam: An expanding war, 1966*. Publ. 1982. 390 pp. 892 refs.

Telfer, Gary L., Lane Rogers, and V. Keith Fleming, Jr. *U.S. Marines in Vietnam: Fighting the North Vietnamese, 1967.* Publ. 1984. 338 pp. 237 refs.

Cosmas, Graham A., and Terrence P. Murray. *U.S. Marines in Vietnam: Vietnamization and redeployment, 1970-1971*, eds. William R. Melton and Jack Shulimson. Publ. 1986. 486 pp. 1,421 refs.

FIELD MANUALS

The following list is not exhaustive, but suggests the range of available materials.

Field Manual 31-15. Operations against irregular forces. Washington, D.C.: Headquarters, Department of the Army, 1961. 48 pp.

Field Manual 31-11 and 31-12 series. Special military occupational specialty manuals and training guides, 1978-1980 and updates.

Field Manual 31-18. Long-range reconnaissance patrol company. Washington, D.C.: Headquarters, Department of the Army, 1968. 30 pp.

Field Manual 31-21. Guerrilla warfare and special operations. Washington, D.C.: Headquarters, Department of the Army, 1961. 259 pp.

Field Manual 31-72. Mountain Operations. Washington, D.C.: Headquarters, Department of the Army, 1959.

Field Manual 33-1. Psychological operations. Washington, D.C.: Headquarters, Department of the Army, 1979. 261 pp.

Field Manual 33-5. Psychological operations. Washington, D.C.: Headquarters, Department of the Army, 1962. 301 pp.

Field Manual 100-2-2. The Soviet army: Specialized warfare and rear area support. Washington, D.C.: Headquarters, Department of the Army, 1984. 103 pp.

Includes material on airborne, heliborne, amphibious and unconventional operations; mountain and desert warfare, and KGB and MVD troops.

Field Manual 100-20. Low-intensity conflict. Washington, D.C.: Headquarters, Department of the Army, 1981. 280 pp.

Fleet Marine Force Manual 8-1. Special operations. Washington, D.C.: Headquarters, United States Marine Corps, 1974. 377 pp.

United States Superintendent of Public Documents. List of field manuals and technical manuals. Washington, D.C.: United States Government Printing Office, 1941; 1951.

Wiseman, John. *Survive safely anywhere: the SAS survival manual*. New York: Crown Publishers, 1986. 288 pp.
A compendium of techniques by a former SAS survival instructor.

(Special thanks are due to John Greenwood, Chief Historian of the Corps of Engineers, who pointed out a wide array of source material; to Will Mahoney and Richard von Doenhoff at the National Archives of the United States for their aid in identifying the main pertinent ranges of documents; and Dean Allard of the U.S. Navy Historical Division.)

9

CRITIQUES

Ahmad, Eqbal. "Revolutionary war and counter-insurgency." In *Revolutionary war: Western response*, eds. David S. Sullivan and Martin J. Sattler. New York: Columbia University Press, 1971, pp. 1-47.

An essay on potential pitfalls in relying on military forces for response in low-intensity conflict. 65 refs.

Ambler, John Stewart. *The French army in politics, 1945-1962*. Bowling Green: Ohio State University Press, 1966. 427 pp.

Substantial discussion and references to *corps d'élite* and political and psychological dimensions of special operations in French colonial wars. 481 refs.

Bergôt, Erwan. *Le dossier rouge: Services secrets contré FLN*. Paris: Bernard Grasset, 1976. 314 pp.

Presents details regarding French covert/special operations against the Algerian nationalists, from the early 1950s to French withdrawal.

Blackstock, Paul W. "Covert military operations." In *Handbook of military institutions*, ed. Roger W. Little. Beverly Hills: Sage, 1971, pp. 455-499.

A rigorous critique written in the final phase of American involvement in Vietnam in the 2nd Indochina War.

----------. *The strategy of subversion: Manipulating the politics of other nations*. Chicago: Quadrangle: 1964. 351 pp.

A sympathetic but cautionary essay on covert operations, on the eve of the 2nd Indochina War; eerily prescient. 404 refs.

----------. "The United States intelligence community and military intervention." In *The Limits of Military Intervention*, ed. Ellen Stern. Beverly Hills: Sage, 1977, pp. 1257-279.

Includes cautions on the tendency of covert operations to exceed initial intentions and bring about "disastrous political consequences," and reviews the "Academic Critics of Covert Operations" in the immediate aftermath of Vietnam. 29 refs.

Bloomfield, Lincoln P., and Amelia C. Leiss. *Controlling small wars: A strategy for the 1980s*. New York: Knopf, 1970. 444 pp.

A proposal in the late phase of the Second Indochina War that American foreign policy abandon counterinsurgency and special operations and rely wholly on diplomacy.

Bradford, Zeb B., and Frederic J. Brown. *The U.S. Army in transition*. Beverly Hills: Sage, 1973. 256 pp.

A post-mortem of the 2nd Indochina War by serving officers (both lieutenant colonels at the time) critical of U.S. military doctrine in the wake of U.S. withdrawal. Rejecting American involvement in large-scale low-intensity operations due to the inability of the U.S. military system to produce forces adequate to such task, the authors praised Special Forces, but saw them as too demanding of resources, given their peripheral role. 6 refs.

Case, Frank B. "Airborne: The tired revolution." *Military Review* (45:8) August, 1965, pp. 86-94

An article written on the eve of the Second Indochina War, as it became apparent that helicopters and the "air-mobile" concept had substantially reduced the role of parachute troops in U.S. Army doctrine and practice.

Clayton, Anthony. *Counterinsurgency in Kenya: A study of military operations against Mau-mau*. Nairobi: Transafrica Publications, 1976. [Transafrica Historical Paper No. 4.] 63 pp.

Fitzgerald, Frances. *Fire in the lake*. Boston: Little Brown, 1972. 491 pp.

Includes a critique of U.S. Special Forces and Provincial Reconnaissance units' terror tactics. 275 refs.

Heggoy, Alf Andrew. *Insurgency and counterinsurgency in Algeria*. Bloomington: Indiana University Press, 1972. 327 pp. 240 refs.

Kelly, George Armstrong. *Lost soldiers: The French army and empire in crisis, 1947-1962*. Cambridge: MIT Press, 1965. 404 pp.

A lucid, detailed exposition of the complex interactions between *corps d'élite*, special operations, psychological warfare, and politics. 163 refs.

Klare, Michael, and Peter Kornbluth, eds. *Low-intensity warfare: Counterinsurgency, proinsurgency and antiterrorism in the '80s*. New York: Pantheon, 1988. 304 pp.

Perrault, Gilles. *Les parachutistes*. Paris: Editions de Seuil, 1961. 191 pp.

A dissection of French *corps d'élites* in Algeria from psychological and sociological perspectives, tracing parallels between such units and the *Waffen SS*. While deeming the *paras* "our S.S.," it also offers extenuating arguments in favor of the French.

Philipps, David Atlee. "Our contras at the Bay of Pigs disaster." *The Boston Globe*, April 10, 1986, p. 15.

A critique of one of the most dramatic failed special operations in American history, by a former senior CIA official, drawing parallels to the Central American situation in early 1986, and arguing that the lack of initial resolve and clear vision by all major actors was the primary cause.

Powers, Thomas. "Lessons of history shadow secret armies." *Houston Chronicle*, August 22, 1985, sec. 1, p. 27.

A critique of American public indifference to the incursions of Soviet-backed regimes in the Cold War, and a catalogue of "melancholy" abandonment of U.S.-backed rebels since 1945 in the Ukraine, Albania, north Burma, Laos, Vietnam, Indonesia, Tibet, Iraq, Angola, and Cuba.

Prados, John. *Presidents' secret wars: CIA and Pentagon covert operations since World War II*. New York: William Morrow, 1986. 480 pp.

A detailed review of U.S. worldwide paramilitary operations/policies, concluding that "they have contributed little to American national security," calling for an end to CIA involvement in the conduct of covert operations, but arguing that problems have stemmed basically from the "imperial presidency" and that "the mere existence of certain types of capabilities can create controversy, even in the absence of any effort to do so"; includes details on command and control structures and key policy documents and an excellent bibliography. 406 refs.

Prouty, Leroy Fletcher. *The secret team: The C.I.A. and its allies in control of the United States and the world*. Englewood Cliffs, N.J.: Prentice-Hall, 1973. 496 pp.

A critical discussion of special operations, especially Chapter 19, "The New Doctrine: Special Forces and the Restorations of the Mutual Security Program." 22 refs.

Schlesinger, Arthur M., Jr. *A thousand days: John F. Kennedy in the White House*. Boston: Houghton-Mifflin, 1965. 1087 pp.

Recollections of the Kennedy administration by a noted historian and advisor, including details on Kennedy's interest in the doctrine and minutiae of special forces and the problems of "overmilitarization" and "over-Americanization" of the U.S. response in counterinsurgency operations.

Talbott, John E. "The myth and reality of the paratrooper in the Algerian War." *Armed Forces and Society* (3:1) 1976, pp. 69-86. 57 refs.

Twohig, J.P.O. "Are commandos really necessary?" *Army Quarterly* (57:1) October, 1948, pp. 86-88.

A critique of an article in the November 1947 *Royal United Services Institute Journal* by General Laycock praiseful of commandos; Twohig argued that commando organization turned away from the original mission of instilling fighting spirit in the army, drained leaders away, had weak logistics due to a regression to quartering, had lack of combat staying power, and wasted training and personnel resources. He concluded calling for fusion of army and Royal Marine special elements, and a solid support base.

10
Popular Images

A vast body of popular culture material on special operations and elite forces has been generated since World War II, and especially since the 2nd Indochina War in the U.S., and since the Falklands War in Britain. Magazines focusing on special units and operations--some conveying an enthusiastic and heroic image, a strong political flavor, hostility toward the media, and so forth--and linked advertising have included:

Against All Odds
Eagle
Gung-ho
International Combat Army
New Breed
New Combat Systems
Soldier Of Fortune
Special Weapons and Tactics

Catalogues featuring special units uniforms, insignia, weapons, and equipment have been advertised in these and other, but not all, military-related publications in the U.S. Paperback novels and such series as Bantam Books' "The Fighting Elite" series have also proliferated. The following list is designed to provide a sense of subjects and approaches and, again, is not exhaustive.

Arnold, Elliott. *The commandos*. New York: Duell, Sloan and Pierce, 1942. 330 pp.
A novel without evidence of much more than popular periodical-based familiarity with the commandos and combined operations.

Berry, John S. *Those gallant men*. Novato: Presidio Press, 1984. 173 pp.
An impressionistic narrative, partly fictionalized, based on the author's experiences as a defense lawyer in the "Green Beret Murder Case."

Berthold, Will. *Brandenburg division*, trans. Alan Neame. London: Anthony Gibbs and Philips, 1961. 215 pp.

Davie, Michael, ed. *The Evelyn Waugh diaries*. London: Weidenfeld, 1976. 814 pp.

Substantial material on Waugh's experiences in a series of British *corps d'élite* in World War II--Royal Marines, Commandos, Royal Horse Guards, Airborne, and the Special Service Brigade. Included in the diary is a "Memorandum on Layforce" covering the diary gap from November 1940 - Spring 1942. Waugh did not "get on" in his various postings, as is evident in his trilogy, and in the first part of *Brideshead Revisited*, written during the war.

Judd, Alan. *A breed of heroes*. London: Hodder and Stoughton, 1981. 288 pp.
A novel portraying the tour of an officer in a British "assault commando" unit in Ulster.

King, David. *The brave and the damned*. New York: Paperback Library, 1966. 285 pp.
A novel based on interviews with Marauder veterans.

Larteguy, Jean. *The centurions*. New York: E.P. Dutton, 1961. 487 pp.

----------. *The praetorians*. New York: E.P. Dutton, 1966. 352 pp.
These novels, made into a film (*The Lost Command*) starring Anthony Quinn and Alain Delon, depict a mythical French airborne unit from Indochina and Dien Bien Phu through to the Algerian revolt, the Battle of Algiers, the coup threat, de Gaulle's coming to power, the humiliation of the *paras*, and the appearance of the OAS--the Secret Army Organization.

----------. *The hounds of hell*, trans. Xan Fielding. New York: E.P. Dutton, 1966. 480 pp.
A novel of the Katanga separatist revolt in the wake of Belgian withdrawal from the Congo, and the role played by Belgian *paras* and foreign mercenaries.

Little, Lloyd. *Parthian shot*. New York: Ivy Books, 1987. 289 pp.
A satirical novel of Vietnam by a special forces medic.

Miller, Kenn. *Tiger the lurp dog*. New York: Ballantine, 1983. 245 pp.
A novel about members of a long range reconnaissance unit in Vietnam.

Padden, Ian. *U.S. air commando: The men, the machines, the battles*. Toronto: Bantam Books, 1985. 133 pp.
A popular history of the special operations group.

Phillips, Gene. *Evelyn Waugh's officers, gentlemen and rogues: The fact behind his fiction*. Chicago: Nelson Hall, 1975. 180 pp.

Reeves, James R., and James C. Taylor. *Covert actions*. New York: Ballantine Books, 1987. 376 pp.
A Vietnam novel with special operations motif.

Rubin, Jonathan. *The barking deer*. New York: George Braziller, 1974.

Sadler, Barry. *Phu-Nham*. New York: Tor Books, 1984. 285 pp.
Sadler, who served in special forces, was well-known in the early 1960s for his popular song, "The Ballad of the Green Berets."

Thompson, Leroy. *Uniforms of the elite forces*. Poole: Blandford Press, 1985. 130 pp.

----------. *U.S. special forces of World War Two*. London: Arms and Armour Press, 1984. [Uniforms Illustrated No. 1.] 68 pp.
Two of many popular illustrated format books appearing in the mid-1980s as nostalgia for Vietnam mounted, and special operations were refurbished in the Reagan administration.

Waugh, Evelyn. *The sword of honour* trilogy.
Novels based on Waugh's experience in the commandos and special operations in the Balkans. (See both Davie and Phillips above.)

White, Kent. *Prairie fire*. Canton, Ohio: Daring Books, 1983. 182 pp.
A novel on rural forces' border-crossing operations in the 2nd Indochina War, by a 5th special forces group veteran.

FILMS:

The following list, again well short of being exhaustive, suggests the popularity of the *corps d'élite* motif in films since World War II:

Apocalypse Now
Battle Cry
The Battle of Algiers
Battleground
A Bridge Too Far
The Bridge on the River Kwai
The Cockleshell Heroes
Commando
Commandos Strike at Dawn
The Dam Busters
Darby's Rangers
D-Day: The Sixth of June
The Devil's Brigade

Force 10 from Navarone
The Green Berets
Gung Ho!
The Guns of Navarone
Heroes of Telemark
Jump Into Hell
The Last Option
The Longest Day
The Lost Command
March or Die
Marine Raiders
Objective Burma
Ocean's 11
Paratrooper
Play Dirty
Private's Progress
Rambo: First Blood Parts I & II
Rommel, the Desert Fox
Tobruk Commando
Uncommon Valor
The Wild Geese

Appendix A: Contemporary Elite Force List by Country*

*Based mainly on data in *The Military Balance*. London: International Institute for Strategic Studies, 1987.

Afghanistan:
- 2 mountain regiments
- 1 commando regiment
- 1 parachute regiment

Algeria:
- 1 airborne special forces brigade
- 4 parachute battalions
- 12 companies of desert troops

Angola:
- 2 "guerrilla force" battalions

Argentina:
- 3 mountain infantry brigades
- 1 airborne brigade
- 1 Presidential Guard regiment (ceremonial)
- Marines - 3,000
- Halcon 8 - army commando counterterror teams
- 4 air force counterinsurgency squadrons

Australia:
- 1 Special Air Service regiment

Austria:
- 2 mountain battalions
- 1 guards battalion
- 1 air-mobile battalion
- *Gendarmerieeinsatz Kommando* - police operations unit - code-named "Cobra"

Belgium:
- 1 paracommando regiment
- 3 battalions of reconnaissance and support units, including *Chasseurs d'Ardennes*
- ESI - *Escadron Special d'Intervention* - Royal Gendarmerie antiterror force

Benin:
- 1 paracommando battalion

Bolivia:
- 1 Presidential Guard infantry regiment
- 1 parachute battalion
- 2 ranger regiments
- 1 marine battalion
- 1 special operations group
- 1 air force counterinsurgency squadron

Brazil:
- 1 airborne brigade - comprised of 3 airborne and 1 special forces battalions
- 2 jungle brigades
- Marines - 15,000 including 1 special operations group
- 3 air force counterinsurgency squadrons
- *Projecto Talon* - special forces antiterror force

Brunei:
- 1 Gurkha battalion

Bulgaria:
- 1 army parachute regiment
- 1 air force parachute regiments

Burkina-Faso:
- 1 airborne battalion
- Honour Guard unit

Burma:
- 1 marine battalion

Burundi:
- 1 commando battalion
- 1 parachute battalion

Cameroon:
- 1 paracommando battalion

Canada:
- 1 airborne regiment
- Canadian Rangers
- Special Service Force - created in 1977 - c. 3500 - infantry, armor, engineers, paratroops, plus support elements; to "be quickly inserted as Canada's initial presence in any national or international theater of operation"
- Emergency Response Teams - Royal Canadian Mounted Police antiterrorist units

Chad:
- Presidential Guard battalion
- 3 Nomad companies
- parachute companies

Chile:
- 1 helicopter-borne ranger unit
- 6 commando battalions
- 4 marine groups, each including 1 commando company and 1 amphibious battalion
- 2 air force counterinsurgency squadrons
- COVEMA - *Commando de Vengadores de los Martires* - covert antiterror force

China (People's Republic):
- 3 air force airborne divisions
- 9 marine regiments and reconnaissance units

Colombia:
- 1 ranger battalion
- 1 parachute battalion
- 1 Presidential Guard battalion
- 2 marine battalions; 3 separate companies and commando units
- 2 air force counterinsurgency squadrons
- *Brigada de Institutos Militar* - anti-guerrilla force

Congo:
- 1 paracommando battalion

Costa Rica:
- Presidential Guard unit
- 1 counterinsurgency battalion

Czechoslovakia:
- 1 airborne regiment

Denmark:
- *Jagerskorpset* - Danish Rangers - created 1961
- *Slaedepatruljen Sirius* - Sledpatrol Sirius - Greenland-based Danish arctic patrol unit - established 1950
- *Sovaernets Fromandskorps* - frogman unit - created 1970

Dominican Republic:
- Presidential Guard battalion
- 1 battalion of naval infantry
- 1 commando unit
- National Police special operations unit

Djibouti:
- 1 commando battalion
- 1 parachute company

Ecuador:
- 1 Presidential Guard squadron - 2 battalions
- 1 special force brigade - 2 battalions
- 1 marine commando battalion
- 1 Puma group - army counterterror unit

Egypt:
- Force 777 - Army antiterror commandos

El Salvador:
- 1 parachute battalion
- 1 naval commando company
- 1 parachute brigade
- 1 Republican Guard brigade
- Orden - *Organizacion Democratica Nacionalista* - anti-guerrilla/terror force

Ethiopia:
- 4 paracommando battalions

Finland:
- *Läskuvarjo Jäakärikoulu* - airborne rangers
- *Osasto Karhu* - Helsinki police anti-terrorist "Bear Unit"

France:
- 2 parachute regiments
- Rapid Action Force components:
- 1 parachute division
- 1 air-portable marine division
- 1 Alpine division
- 1 light armored division
- 1 air-mobile division
- Foreign Legion - 1 armored, 1 parachute, 4 infantry, 2 engineer regiments - 8,300
- 2 helicopter assault squadrons
- naval commando battalion
- *Groupe d'Intervention de la Gendarmerie Nationale* - antiterrorist response force - officially less than a hundred members, with about a third on alert - also called *Gigène*
- *Groupement d'Intervention* - French overseas rapid-deployment force, various forms, 1960s-1980s - airborne, Foreign Legion units

Gabon:
- Presidential Guard battalion

Germany, East (German Democratic Republic):
- Ministry of State Security Guard regiment
- 2 amphibious regiments
- 1 special forces brigade
- 1 airborne battalion

- 8 naval reconnaissance boat groups
- 40th (Willi Sänger) Parachute Battalion
- *Wachregiment* - special unit under control of Ministry of State Security
- *Aufklarungstrüppe* - penetration forces

Germany, West (Federal Republic of Germany):

- 1 mountain division
- 1 mountain brigade
- 1 airborne division
- 3 airborne brigades
- *Luftwaffe* special operations wing
- border police, including *Grenzschützgruppe 9* - "Border Security Group" - special German border guard/army unit - formed in mid-1970s after Bavarian police botched Munich Olympic rescue attempt - GSG-9 carried out Operation MAGIC FIRE, retaking of a hijacked Lufthansa airliner at Mogadishu, Somalia in 1973
- *Fernspay* Companies - special advance reconnaissance teams attached to Corps Headquarters of the *Bundeswehr*
- *Spezialeinsatzkommandos* - based on GSG 9 - platoon-sized antiterror teams under states of Federal Republic of Germany - formed in mid-1970s after Munich debacle for quick local response - low casualties in various operations

Ghana:

- 1 parachute battalion

Greece:

- 1 paracommando division - comprised of 1 parachute, 1 commando, and 1 marine regiment
- 3 commando battalions
- 3 light army battalions
- DEA - *Dimoria Eidikon Apostolon* - Athens police special antiterrorist platoon

Guatemala:

- 1 Presidential Guard brigade - 2 battalions
- 1 special forces brigade - 2 battalions
- 1 air force counterinsurgency squadron

Guinea:

- 1 commando battalion
- 1 special forces battalion

Guyana:

- 1 guard battalion
- 1 special forces battalion

Haiti:

- Presidential Guard
- 1 infantry battalion
- 1 amphibious squadron
- 1 special forces battalion
- *Leopards* - elite force and "palace guard;" trained by U.S. advisors - 1980s bodyguard
- Night and Jungle Special Troops

Honduras:

- 1 special forces battalion
- 1 Presidential Guard company - ceremonial
- 1 air force counterinsurgency squadron

Hungary:

- 1 airborne battalion

India:

- 7 mountain divisions - 3 brigades each
- Presidential cavalry escort
- 1 paracommando brigade - 9 battalions
- 10 Gurkha battalions
- Indo-Tibetan border police, c. 12,000

Iran:

- 1 airborne brigade
- 1 special forces division - comprised 4 brigades
- 3 marine battalions
- elite components within Revolutionary Guard Corps

Indonesia:

- 2 airborne brigades of 3 battalions each
- 4 independent airborne battalions
- 4 special warfare groups
- 5 marine regiments
- 5 counterinsurgency battalions
- 1 air force counterinsurgency squadron
- *Resimen Para Kommando Angkatan Darat* - paracommando assault regiment
- Detachment 81 - special commando unit

India:

- Presidential Guard force - includes 1 Commando brigade
- 2 special forces divisions - includes 6 brigades

Iran:

- *Pasadaran* - Revolutionary Guards - includes elite elements - details of organization and missions uncertain

Ireland:

- 1 ranger company

Israel:

- 7 parachute battalions
- Golani brigade
- 1 naval commando force - 300
- *Sayaret* - designation of various special units: *Sayaret Almond* - desert motorized patrol and raid force under Southern Command; *Sayaret Carob* - anti-infiltration force under Western-Central Command; *Sayaret Egoz* - mountain unit under Northern Command; *Sayaret Matkal* - counter-terror force, 1970s-80s - SAS-like military unit under General Staff - rapid response and undercover roles - worldwide capabilities
- *Mossad* - *Mossad el Aliyah Beth* - largest Israeli intelligence service - includes special operations elements - created 1952

Italy:

- 5 *Alpini* brigades
- 1 airborne brigade
- 2 amphibious battalions
- 3 marine groups - 1 underwater raiding, 1 infantry, 1 logistical support
- 1 *carabinieri* airborne battalion
- GIS - *Gruppo Speziale Intervenzione* (Special Intervention Group); a c. 50 man rapid response team raised from the *Carabinieri* Parachute Battalion--also called "leather-heads" due to wearing of face masks
- NOCS - *Nucleo Operativo Centrale di Sicurezza* - police commando unit of c. 50; formed 1978; freed General Dozier in February 1982
- Lagunaria - Italian forces trained and equipped to negotiate swamps and shallow lagoons near Venice and in the Po Valley and adjacent river deltas
- *Incursori* - underwater special operations penetration units
- San Marco Marines - amphibious elite unit

Ivory Coast:

- 1 parachute company
- 1 Presidential Guard brigade - c. 1,000

Japan:

- 1 airborne brigade

Jordon:

- 1 Royal Guards brigade
- 1 special forces brigade - comprised of 3 airborne battalions
- 101st Special Forces Battalion - antiterrorist force

Kenya:

- 1 parachute battalion

Korea, North (Demoncratic Republic of Korea):
- 25 special operations brigades - c. 100,000
- "Bureau of Reconnaissance Special Forces"; includes 3 commando, 4 reconnaissance, 1 river crossing; and 3 amphibious brigades
- 5 airborne battalions

Korea, South (Republic of Korea):
- 7 special warfare brigades
- 2 divisions and 1 brigade of marines; c. 23,000

Kuwait:
- Palace Guard unit

Liberia:
- Executive Mansion Guard battalion
- air reconnaissance battalion - c. 250

Libya:
- 14 paracommando battalions

Malaysia:
- 1 special service regiment - 3 battalions

Mexico:
- Presidential Guard brigade
- 13 marine groups - 6,500
- 6 air force counterinsurgency squadrons

Morocco:
- 1 parachute brigade
- 1 Royal Guard battalion
- 1 mountain battalion
- 4 commando battalions
- 4 camel corps battalions
- 2 desert cavalry battalions
- 1 naval infantry battalions

Nepal:
- Royal Guard brigade

Netherlands:
- 2 amphibious commando groups
- 1 mountain/arctic warfare company
- 1 amphibious Commando group reserve
- marines - 2,500

New Zealand:
- 1 special air service squadron
- 1 air force counterinsurgency squadron

Niger:

- 1 parachute company

Nigeria:

- Composite division - 1 airborne, 1 air-transportable and 1 amphibious brigade

Norway:

- Royal Guards battalion
- *Jeger* - independent ranger platoons
- *Beredskaptrop* - national police alert force

Oman:

- 1 Royal Guard brigade
- 1 special forces regiment

Pakistan:

- 1 special services group of 3 battalions

Panama:

- 1 airborne company
- 1 Special Forces company

Paraguay:

- 1 Presidential Escort regiment
- commando battalion

Peru:

- 1 airborne brigade of 3 battalions
- 1 jungle battalion
- Marines - 3,500

Philippines:

- 1 ranger regiment of 2 scout/ranger battalions and 1 mountain battalion
- 1 Presidential Security command of 6 battalions
- 3 marine brigades - 9,500
- AVESCOM - Aviation Security Commando force
- LRF - Philippine Constabulary Light Reaction Forces

Poland:

- 1 airborne division
- 1 amphibious assault division
- 7th Lujycka Naval Assault Division - unit of the Polish Army - major Warsaw Pact amphibious unit in the Baltic

Portugal:

- 1 special operations brigade, comprised of 1 commando regiment and 4 support battalions
- 3 battalions of Marines - 2,600, including infantry and police units
- 1 air force parachute brigade of 3 battalions

Qatar:
- Royal Guard regiment

Rumania:
- 3 mountain brigades
- 2 airborne regiments

Rwanda:
- 1 commando battalion

Saudi Arabia:
- 1 airborne brigade, comprised of 2 parachute battalions and 3 special forces companies
- Royal Guard regiment
- Cavalry Squadron - ceremonial
- Ministry of Interior helicopter-borne counter-terror unit

Senegambia:
- Presidential Guard unit - mounted
- 2 parachute companies

Singapore:
- 1 commando battalion, active; 1 reserve
- 1 Gurkha battalion
- 3 air force counterinsurgency squadron

South Africa:
- 1 special reconnaissance regiment
- 1 parachute brigade of 3 battalions
- 1 marine battalion

Southwest Africa:
- 1 parachute battalion

Spain:
- 2 mountain divisions - 2 brigades each
- Royal Guard regiment - all services elements
- 1 air-portable division
- 1 parachute brigade
- Spanish Foreign Legion - 3 regiments, support units, and special operations group
- marines - 11,500
- *Policia Nacional*-Special Operations Commando Group (GEO)
- *Guardia Civil*-antiterrorist group (UAR)
- *Grupo Especiale Para Los Operaciones* - Spanish counterterror force
- Spanish Foreign Legion - founded in 1921 by Colonels Millan Astray and Franco for Moroccan service after Anual disaster - fought against republic in Spanish Civil War - c. 7,000 men

Sri Lanka:
- 1 special forces task force - c. 4,000
- special forces regiment

Sudan:
- 1 Republican Guard brigade
- 1 parachute brigade

Suriname:
- 1 commando company
- 1 guard battalion

Sweden:
- *Basjägare* - air force ranger security force
- *Faltjägare* - forest and arctic rangers
- *Fallskarmsjägere* - parachute-rangers - 1 company
- *Kustjägare* - coastal/marine rangers - 1 company
- coast artillery regiments - equivalent to marines

Switzerland:
- 3 mountain infantry division
- 1 mountain infantry regiment

Syria:
- 1 Special Forces Division of 9 paracommando brigades
- Republican Guard unit
- Desert Guard - c. 1,500

Taiwan:
- 2 airborne brigades
- special forces - numbers uncertain
- 3 marine divisions - c. 40,000

Thailand:
- Royal Guard
- 2 special forces divisions
- 8 marine battalions - c. 20,000
- special action forces - 3,000
- rangers - 13,000
- Border Patrol Police
- 7 air force counterinsurgency squadrons

Togo:
- 1 paracommando battalion
- 1 Presidential Guard regiment, including 1 commando battalion and 1 guard battalion

Tunisia:
- 1 paracommando brigade
- 1 Sahara brigade
- *Groupement de Commando - Garde Nationale* antiterror force

Turkey:

- 1 parachute brigade
- 1 commando brigade
- 1 marine brigade - 4,000 - 3 infantry and 1 artillery battalions, plus support elements

Union of Soviet Socialist Republics:

- Kremlin Guard (KGB)
- Special Guard and signals units (KGB)
- naval infantry - 18,000
- 7 airborne divisions
- 16 brigades and 3 separate regiments of *spetsnaz* (special forces)
- 4 naval *spetsnaz* brigades, 1 per major fleet command
- elements of KGB and MVD special fully armed combat troop units, out of 670,000 total, including Border Guards; exact numbers and allocations unknown
- *Okhotriki* - designation of Soviet *spetsnaz* "hunter" units in Siberia
- *Reydoviki* - designation of Soviet *spetsnaz* raider units in Siberia
- *Vysotniki* - small Soviet navy *spetsnaz* detachments
- *Spetsnaz* - Soviet security organs' special units - estimated at 15-30,000 - played key spearhead roles in Prague airport seizure, 1968, and in occupying Kabul, 1979 - believed to "scope" possible operational areas as tourists - links to Special Troops of KGB and MVD and to terrorist training purported but unclear

United Arab Emirates:

- 1 Royal Guard battalion

United Kingdom:

- Household Division - 5 battalions infantry, 2 cavalry regiments
- 3 parachute battalions
- 6 Gurkha battalions
- 1 Special Air Service regiment
- 1 Royal Marine Commando Brigade
- 1 Royal Navy Special Boat Squadron
- 3 helicopter assault squadrons
- Greenjackets
- Special Boat Section - Royal Navy special unit counterpart to SAS
- C-13 - police antiterrorist squad - "blue berets"
- Special Branch - c. 400 man counterintelligence unit orginally raised in late 19th century to deal with Irish insurgents - active in Imperial policing

United States of America:

Non-military:
- HRT - FBI Hostage Response Team
- NEST - Department of Energy Nuclear Emergency Search Teams

Army:
- 1 airborne division
- 1 ranger regiment of 3 battalions
- 8 Special Forces groups - 4 Active Army; 2 National Guard; 2 Army Reserve
- Delta Force with Air Force attack helicopter and transport group
- 1 mountain infantry battalion
- 2 army special operations aviation battalions
- 1 Arctic scout unit
- 4 psychological warfare groups (13 battalions)
- Department of Energy emergency teams
- FBI counterterror teams

Navy:
- 4 special warfare groups, including 6 SEAL teams; 2 SEAL delivery teams; 2 special boat squadrons; 3 nuclear transporter submarines; underwater demolition units
- force reconnaissance sub-elements in Marine Amphibious Forces
- 2 light assault squadrons

Air Force:
- 5 special operations squadrons with specialized aircraft
- 2 reserve groups

Marines:
- MAU (SOC) - Marine Amphibious Unit (Special Operations Capable) - c. 3,000 men each - one each in Atlantic and Pacific

Uruguay:
- Presidential Escort regiment
- naval infantry battalion
- Metropolitan Guard battalion
- Republican Guard battalion
- 2 air force counterinsurgency squadrons

Venezuela:
- 2 army airborne battalions
- 1 air force airborne battalion
- marines - 4,500
- 1 air force counterinsurgency squadron

Vietnam:
- 10 marine brigades
- naval infantry - c. 27,000

Yemen, North (Yemen Arab Republic):
- 1 special forces brigade
- 1 paracommando brigade
- Central Guard Force

Yugoslavia:
- 3 mountain brigades
- 2 marine brigades, of 2-battalion regiments each
- 1 airborne brigade

Zaire:
- 1 special forces division
- 1 Presidential Guard brigade
- 2 parachute brigades
- 1 special forces brigade

Zimbabwe:
- Presidential Guard unit
- 3 guards battalion
- 1 commando battalion
- 2 parachute battalions
- 1 air force counterinsurgency/reconnaissance squadron

Appendix B: Some Elite Forces Formed Since 1939

African Coastal Flotilla - British naval special unit - World War II

Airborne units - U.S.S.R., Germany, France, Japan, Britain, U.S., Israel, Belgium, Republic of South Vietnam et al.

Air Studies Group - 90th Special Operations Wing - U.S. Special Forces and first-line South Vietnamese Air Force units - 2nd Indochina War

Alamo Scouts - U.S. special spearhead unit in MacArthur's Southwest Pacific Command, 1943-45

Air Commandos - U.S. Army Air Forces and U.S. Air Force- first version, Special Operations Wing, formed in Burma, under John Alison and Phil Cochran in World War II - active in Vietnam, in "Jungle Jim" and following phases - currently called Special Operations Groups

American Volunteer Group - "Flying Tigers" - raised in U.S. from U.S. military fliers by Nationalist Chinese, late 1941 - fought briefly in Burma and disbanded, 1942 - founded by Claire Chennault

Apache Force - Special Forces teams which "trained" arriving U.S. forces in Vietnam and accompanied in early contact - also served as reserve and reaction force

Arab Legion - formed by Sir John Glubb (Glubb Pasha) to fight Saudi raiders and serve as Trans-Jordan constabulary in early 1920s

Australian Independent Companies - special scout-raiding units formed in Far East 1939 - active throughout Pacific War

Aviation Task Force 160 - U.S. Army "Night Stalker" helicopter force in Persian Gulf, late 1980s

Beach Commandos - an upgrading of British beach controllers, after Dieppe

Beach Jumpers - U.S. Marine Corps special advanced teams in early Pacific War assualt landings - disbanded along with other Marine special units in 1944

Black Commandos - French units formed in Algeria to move among and work with the Arab population, eschewing heavy firepower, brutality, and racism - some elements also called Nomad Commandos

Blue Light - code name for U.S. special antiterrorist force formed in late 1970s from Special Forces

Bomb Damage Assessment Units - special units under SOG in 2nd Indochina War

Bomb Disposal Units - formed by several countries in World War II - rarified elite force requiring strength, stamina, coolness, mechanical and problem-solving skill, and raw courage

Boom Clearance Party - British naval special unit - World War II

Brandenburgers - formed under Division II of the German *Abwehr* (Military Intelligence service) for sabotage and special duties just prior to World War II - initially company-sized, with mainly Balts and *Volksdeutsche* (foreign-born German-speaking) personnel - ZbV800 = *Zur Besondern Verwendung* ("for special use")

Burma Defense Army cadres - c. 2800 - selected from c. 8,000 of Burma Independence Army by Japanese to attend special training centers - later led Burma National Army - many prominent in post-World War II Burma

Burma Intelligence Corps - provided guides and interpreters for British forward patrols in World War II

Cao Dai - Vietnamese religious sect north of Saigon - constituted a division in South Vietnamese Army and a preeminent mercenary force used by Special Forces in the CIDG

Carpetbaggers - 8th Air Force units supporting special/partisan operations in Europe in northwest Europe in World War II

Charioteers - Royal Navy special "human torpedo" unit, World War II

Chindits - Long Range Penetration Force - name derived from corruption of "Chin-the," Burmese devil-lion temple guardian statues - columns of Gurkha, West African, and British forces supplied from the air to raid in strength deep behind Japanese lines in Burma, 1943-44 - followed doctrine of Orde Wingate - also called "Wingate's Raiders" - some glider-borne in 1944

Civilian Airborne Rangers - CIA-raised special operations forces in 2nd Indochina War

Combat air crews - including various "elites within elites" - e.g., Pathfinders, Black Sheep Squadron, naval aviators, "aces," veterans of the "Hump" (trans. Himalayan supply route to China)

Combined Operations Assault Pilotage Parties - COPP - cover title for British underwater/amphibious sabotage commando force, World War II

Cossack Division - Soviet army World War II elite force

Commando Ponchardier - French naval commando unit in Indochina War

Commandos - British raiding force formed in 1940 - initially raised from Army units under Combined Operations - all volunteer, severe training, high casualties - put under Royal Marines in 1943 - conducted several dozen raids of various sizes - served in Korea, Falklands and various British limited warfare operations since 1945

Commachio Company - Royal Marine Commando special unit of 1980s - named for an action in Italy in World War II - trained to protect Britain's offshore oil infrastructure against hijacking and sabotage - a sub-unit stationed at the Faslane nuclear submarine port

Delta Force - special U.S.-Vietnamese command including Vietnamese Rangers, reconnaissance teams, "roadrunner" infiltration teams, and Nung headquarters guards formed in U.S. Army in mid-1970s along lines of British SAS - commanded by Colonel Charles Beckwith in early phase - involved in aborted Iran hostage rescue in April 1980, Operation EAGLE CLAW, and at *Achille Lauro* standoff on Sicily - wracked by fiscal scandals in mid-1980s

Desoto Patrols - raids into North Vietnam in early 1960s

Detachment 385 - British naval special unit in World War II

Donkey teams - U.S.-South Korean partisan teams in Korean War

Dzerzhinski Division - Soviet NKVD division in World War II - predecessor of KGB armed forces unit

Eagle Flight - Special Forces-led rapid-response and reserve platoons of *montagnards* in 2nd Vietnam War

Einheit Steilau - the sub-unit of American-English speaking infiltrators under Skorzeny's overall command in the Battle of the Bulge, 1944

Engineer Special Brigades - U.S. multi-service advanced landing detachments in WWII amphibious operations

Experimental Combined Action Platoons - specially selected Marines with a medical corpsman living in villages to provide security and civic aid in I Corps area - most successful of U.S. counterinsurgency programs

15th Motor Gunboat Flotilla - Royal Navy special unit - landed and took off agents - Europe - World War II

1st U.S.-Canadian Special Service Brigade - raised to launch major attack on. Norwegian industry - diverted to light infantry roles in Aleutians, Italy, and France - commanded by Colonel Robert D. Frederick - an "ancestor" unit in official lineage of U.S. Special Forces

Fiji Scouts - 1 Commando Fiji Guerrillas - under British command World War II

Folgore Division - Italian airborne division - originally formed in World War II

Force 101 - OSS-led Burmese tribesmen groups in World War II

Force 136 - SOE's Indian Mission title after 16 March 1944 - oversaw most major British special operations in Southeast Asia

Force "J" - a permanent at-the-ready assault force formed by Combined Operations after Dieppe, where lack of a deployable reserve was a major problem

Force Recon units - U.S.M.C. deep penetration forces first in 2nd Indochina War - currently the U.S. Marine Corps' principal special unit

Force 777 - Egyptian counterterror commandos

Force Viper - British commando force in Burma - World War II

Force "X" - Philippine special force, Huk rebellion, late 1940s-1955 - deception and ambush operations - also in 2nd Indochina War

Force 17 - elite commando element of Palestinian *Al Fatah* - about 2,000 in detachment throughout Europe - originally small bodyguard force protecting Yassar Arafat - sub-elements trained in underwater warfare

Foreign Operating Group - created mid-1980 - later became Intelligence Support Activity

Forward Air Controllers - FACs - U.S. Air Force and Army light aircraft pilots calling in fire, air strikes and advising nearby units - high risk and high casualties

Four-Excellence Companies - elite Chinese Communist units

Friedenthal Kommando - a special *SS* unit under Skorzeny's command

French Foreign Legion - formed in 1830s - an active *corps d'élite* in French colonial wars in Africa, Asia, and Mexico including the 1st Indochina War and Algeria, and - the World Wars - mainly foreign recruits

Gamma - code name for U.S. Special Forces border-crossing Detachment B-57 formed elite force in late 1969 in Vietnam

GCMA - Composite Airborne Commando Groups - c. 20,000 Vietnamese and a small French core organized under by Roger Trinquier in - 1st Indochina War - model for U.S. Special Forces CDIG units

Guards Divisions - Red Army - World War II combat-awarded title

Grey's Scouts - a light cavalry force of about 300 which tracked and patrolled during the Rhodesian insurgency - about half Europeans and half black Africans

Ground Studies Group - U.S. Army Special Forces and Vietnamese units under SOG control in 2nd Indochina War

Groupement Mixte d'Intervention - French-Vietnamese rapid-reaction forces forces in 1st Indochina War

GSI (k) - SOE India Mission title from 1941 to March 1944

Hierchos Lochos - Sacred Company/Squadron - elite force of ex-officers formed after Greek collapse in 1941 - fought under British command until 1944 - cadre of anti-Communist effort, 1947-52

Hoa Hao - Vietnamese Buddhist sect in Mekong Delta - c. 1 million - CIA-linked from early 1960s - used by Special Forces as camp defenders

I-Go Kimmutai - First Task Force - Japanese-led special units of Indonesian volunteers selected from Bogor training center World War II - unter Japanese 16th Army in 1945 - cadre for guerrillas - wore civilian clothes - trained in ambush tactics

Independent Companies - British Army light raiding units formed from Territorial Army early in WWII - some members joined Commandos in 1940

Intelligence Assault Unit - British naval special unit - World War II

Interservice Liaison Department - cover name for SIS (British Secret Intelligence Service) paramilitary operations in Far East, WWII - not under SOE

Interservice Sea Reconnaissance Unit - 40-man "natatorial" British Navy unit trained in Bahamas and California in World War II - aided XIVth Army in Burma, most notably in crossing the Irrawaddy

Jagdverband 44 - Luftwaffe "squadron of aces" - formed in February 1945 - flew jets under command of Adolf Galland - initial successes with rockets against bombers neutralized by Allied hits on base fields

Jagdverbande - anti-partisan "hunting bands" formed by Otto Skorzeny

Jarai Rangers - U.S.-led special unit in 2nd Indochina War

Jedburgh Teams - small detachments of British, French, and American members - coordinated French resistance organization, supply, and operations - controlled by SOE and OSS - active 1944-45 in France

"Jungle Jim" - U.S. Air Force air commandos in Vietnam in early 1960s - under name of 4400th Combat Crew Training Squadron - part of U.S. assistance efforts code-named *Farmgate*

Kampfgeschwader 200 - unit of Luftwaffe that carried out special operations - used Allied aircraft

Kampftrüppen - German *Abwehr* units formed just before World War II - precursors to Brandenburgers - main roles critical point seizure and blocking defensive demolition

K Flotillas - World War II German naval special units - (Klein Kampf Verbande) - formed in reaction to British underwater special units - used mainly against Anglo-American targets

Kit Carsons - North Vietnamese and Viet Cong defectors under Chieu Hoy ("open arms") program who aided U.S. and Allied forces in standard and special operations

Konoye Shidan - Japanese Imperial Guards Division - World War II

Kurfürst Regiment - successor unit to Brandenburgers in World War II when the latter were expanded into a division

Lehr Regiment ZbV800 - cover designation for Brandenburgers

Leibstandarte regiment - Hitler's SS bodyguard regiment

LLDB - South Vietnamese Special Forces regulars - known as 1st Observation Battalion "palace guard," 1956-63 - designated Lac Luong Dac Biet and sent to field operations, 1963 - also known as VNSF - Vietnamese Special Forces

LRRP - Long Range Reconnaissance Patrols - "LURPS" - various types of U.S. volunteer teams ranging in suspected or known Viet Cong areas in Vietnam - under Special Forces and other units received special intense training from Special Forces, Rangers or in "recondo" schools - usually within range of main force artillery

Long Range Desert Group - British deep penetration, surveillance and sabotage motorized unit in North Africa, 1941-43

Malayan Scouts - raised by British in Malayan Emergency in the 1950s - included Commonwealth volunteers

Merrill's Marauders - 5307th Composite unit (Provisional) in Burma, 1943-44 - U.S. counterpart to Chindits - core of Galahad Force - remnants later folded into 475th Infantry Regiment, in MARS Force offical organizational "ancestor" of 75th Infantry Regiment, current Rangers' parent unit

Mike Force - Mobile Strike Force Commands - MSFC - formed June 1965 in Vietnam - one per Corps area and one under SOG - included reaction, deep reconnaissance, and larger support units

Mobile Flotation Unit - British naval special unit - World War II

Morskaya pekhota - Soviet naval infantry; active in World War II; deactivated 1947; reactivated in late 1950s or early 1960s

Mountain troops, including Italian *Alpini*, French *Chasseurs Alpin*, U.S. 10th Mountain Division, German and Austrian *gebirgsjäger*, and Indian mountain divisions

MTTs - U.S. Special Forces Mobile Training Teams - 2nd Indochina War - trained local and special Vietnamese forces - also some U.S. and Allied forces

91st Airborne Ranger Battalion - special unit under Delta in 2nd Indochina War

NKVD combat units - World War II - 1st and 21st NKVD Divisions - Special Division in *VgK Stavka* (Soviet High Command) reserve in 1943 - 10th, 12th, 90th Border Guard Regiments - also OOs and *zagraditelneye otradyi*

Nomad Commandos - see Black Commandos

North Vietnamese Sapper Units

Nung mercenaries - ethnic Chinese who served as Special Forces' bodyguards - prominent in special operations in Vietnam

Omega - code-name for Special Forces field intelligence team - Detachment B-60 in 2nd Indochina War

Operations Centre Khartoum - British special force in Sudan, World War II

Osobyi Gardyeski Batalion Minyorev - Special Guard Assault Battalions - in Soviet NKVD in World War II

Osobyi Otdel - "OOs" - special small counterintelligence units of NKVD in World War II

Panzer Brigade 150 - cover name for Skorzeny's special unit in the Battle of the Bulge

Path Finder Force - Royal Air Force target-marking force of crack navigators, WWII - officially No. 8 Group - under Air Vice Marshal Don Bennett

Pathfinders - 2nd Indochina War - "Black Hats" - Army, Air Force, and Marine special teams who preceded parachute/air landing in clearing zones and navigation systems

Philippine Scouts - 12,000 man elite cadre of US-raised Philippine forces in late 1941-early 1942; include 43rd, 45th and 57th Infantry Regiments and 26th Cavalry Regiment; suffered almost 50% losses in campaign, and many more in captivity

Photo-Reconnaissance Unit pilots - PRU - flew unarmed aircraft to take photos often under heavy fire and at low-level

Popski's Private Army - small reconnaissance/raiding/deception force raised by Vladmir Peniakoff in North Africa, World War II

Provincial Reconnaissance Units - airborne-qualified irregular teams in Vietnam - late 1960s

Q Patrols - British Special Branch/Turkish police teams - on Cyprus in the late 1950s

Queen's Cobras - Royal Thai regiment in Vietnam 1967-68 in III Corps area with U.S. 9th Infantry Division units

Raider and Parachute Battalion, WWII U.S. Marine Corps - formed in Pacific as counterparts to the British Commandos - paratroops did not jump in combat - used as raiding and reconnaissance force

Rangers - U.S. Army force raised in Northern Ireland as U.S. counterpart to commandos - fought mainly as battalions in Europe and Pacific, 1942-45 - disbanded after WWII; Ranger companies in U.S. divisions in Korea - reformed as elements of 75th Infantry Regiment in 1968 - variants in other armies, e.g., Vietnam, Philippines

Rhale Rangers - U.S.-led special unit in 2nd Indochina War

Royal Marine Boom Patrol Detachment ("Cockleshell Heroes") - cover name for British special unit using various exotic small craft - light sabotage and reconnaissance canoe force in WWII

Sa'aqa - Egyptian antiterror "Thunderbolt" commandos - 1970s - replaced by Force 777

Scout-Rangers - Philippine Army - Huk rebellion - early 1950s

Sea Reconnaissance Unit - British naval special unit - World War II

SEALS - U.S. Navy Sea-Air-Land special force units - created in 1962 for sabotage, infiltration reconnaissance, civic aid - evolved from U.S. Navy underwater demolition teams - worked in Vietnam and Grenada with army and Air special units - duties in Vietnam included rescue of prisoners of war in early 1970

Seaspray - Central America-based intelligence gathering unit, in the late 1980s

75th Infantry Regiment - official lineage parent unit of U.S. Army Rangers - descended from 475th Infantry Regiment

Selous Scouts - Rhodesian light force 1960s-70s- named after famous hunter

Shin' Eitai - honor guard formed by Japanese from Filipino volunteers in World War II to protect puppet officials

Shotgun Commando Groups - Greek Cypriot EDKA underground strike force in the late 1950s

Sigma - Special Force field intelligence Detachment B-56 in Vietnam - late 1960s, early 1970s

610 Reconnaissance Commando - French special operations unit during first Indochina War

6004th Air Intelligence Service Squadron - controller of special and partisan operations in the Korean War, 1951-53

SLAM Companies - "Search-Locate-Annihilate-Monitor" - U.S.-Vietnamese special units made up of Hatchet/Hornet platoons - 2nd Indochina War

SOE - Special Operations Executive - British main command in World War II overseeing espionage, sabotage and Special Operations

Sonderkommando Elbe - Luftwaffe "storm detachment" trained to ram Allied bombers in late 1944 - c. 700 volunteers - suffered heavy losses - failed to achieve dramatic impact when used en masse in early 1945 - survivors held in reserve in Austria till war's end

Special Action Unit Number One - special operations unit under 5th U.S. Air Force in Korea, 1951

Special Air Service - founded by David Stirling in North Africa in 1971 - raided German rear areas - branched into French, British, and Belgian battalions under SOE, 1942-45 - variants in U.K., Belgium, Australia, South Africa, and Rhodesia after WWII - special operations and antiterror roles since WWII in Malaya, Kenya, Cyprus, Aden, Oman, Borneo, Sarawak, Ulster, and in Falklands War - some Rhodesian SAS veterans went to South African SAS after disbandment of former in 1980 - Australian SAS in Vietnam SOG 1960s-70s

Special Forces - formed in U.S. Army in 1952 as cadre for guerrilla force a la the wartime SAS - expanded under Kennedy administration - converted to counterinsurgency and covert operations in Vietnam War - cut back in the late 1960s - rebuilt in Reagan administration

Special Service Brigades - various Commando parent organizations and HQs in WWII - also trained other elements

Spike Teams - 2-3 U.S. and 9 Vietnamese special forces units - 2nd Indochina War - reconnaissance teams operating in Vietnam under SOG - late 1960s in Vietnam

SS Parachute Battalion - under Otto Skorzeny's command - numbered "600" at end of World War II

Studies and Observation Group - formed 1964 as major command over special operations in Vietnam - succeeded Saigon Military Mission under Lansdale

Task Force 116 - floating headquarters for U.S. Navy SEAL teams in the Mekong Delta - 1st Indochina War

Task Force 160 - U.S. covert operations unit, 1980s

30 AU - Royal Navy/Commando Assault Unit - World War II - included Forward Interrogation Unit

Tokumu Kikan - "special duty agencies" - Japanese intelligence special teams in Burma in World War II - *F (Fujihara) Kikan* - active in forming first Indian National Army from Indian POWs in 1941-42; *Iwakuro Kikan* - succeeded *F Kikan; Mihima Kikan* - temporary army-navy project

Tunnel Rats - volunteers in U.S. and South Vietnamese forces who entered Vietcong/North Vietnamese tunnel complexes during the 2nd Indochina War

Underwater Demolition Teams - U.S. Navy - created in World War II - still in existence

Unit 269 - Israeli special operations units formed from paratroop unit in 1950s

UNPIGK - United Nations Partisan Infantry Group Korea - mainly Koreans, led by and linked with U.S. covert elements, 1951-53

USAFIP - U.S. armed forces in Philippines; shadow headquarters of U.S. and loyal Filipino guerrillas and informants during Japanese occupation, 1942-45; ultimately encompassed five regiments and c. 20,000 personnel

V Force - British deep penetration fighting and intelligence patrols - Southeast Asia, World War II

Vietnamese (South) Marine Corps

Walli II - sabotage commando element of the German Abwehr/SD - efforts directed by Walter Schellenberg on the Eastern Front

White Tigers - U.S.-South Korean partisan teams, 1951-53

Wolfpack - Koreans raised under U.S. leaders in the Korean War for infiltration - later involved in special operations

Yasu Butai - Japanese special unit in Southeast Asia - World War II

Yuge Kitai - Japanese special units in Indonesia - World War II - wore civilian clothes - moved among population to gather intelligence

Z Force - armed patrols providing command intelligence up to 60 miles behind enemy lines in Burma - controlled by GSI (K) Force 136

Z Special Force - Australian variant of Z Force

Zagraditelnye Otryadi - NKVD holding/blocking detachments in early operations in World War II

Zanchi Yugekitai - Indonesian stay-behind commando force under Japanese officers raised at end of World War II to thwart Dutch reoccupation

Appendix C: Major Counterterrorist Special Operations Since World War II

Battle of Algiers - January-March 1957

French 10th Airborne Division and other security forces halted Algerian FLN bombing attacks through various tactics including torture, leading to a major alienation of the Arab population and polarization in France with major long-terms consequences - seen as a classic case of winning battles and losing the war

Beirut Airport - 28 December 1968

In an hour-long helicopter-borne raid, a forty-man Israeli *Sayaret Matkal* team headed by Brigadier General Rafael Eytan blew up thirteen Lebanese-owned airliners, retaliating for an attack on an Israeli airliner in Athens on the 26th. France cut off aid to Israel and substantial publicity was focussed on the Palestinians

Lod Airport - 9 May 1972

The first successful special operation against a hi-jacked airliner. Two Black September terrorists seized a Belgian airliner in flight from Brussels to Lod, demanding the release of 371 Arab prisoners in Israeli custody. While negotiations were arranged through Red Cross channels, a *Sayaret Matkal* team rehearsed an attack on a Boeing 707 in a hangar close-by. When, dressed as mechanics, the Israelis broke in, the terrorists and a woman passenger were killed. The Red Cross protested having been used to cover the attack.

Beirut - 9-10 April 1973

About thirty Israeli *Sayaret Matkal* raiders landed from the sea and attacked the Black September hierarchy in Beirut. Three Palestinians and two Israelis were killed; supplies destroyed; records captured; extensive civilian casualties. *Mossad (Mossad Le Aliyah Beth* - Institution for Intelligence and Special Services) agents were also involved. Some raiders left by sea, others in cars, helicopters dropping spikes to impede pursuers. Major outrage in the Middle East and demonstrations sympathetic to the Palestinians followed.

Lillehammer - 21 July 1973

Two Israeli *Mossad* agents shot a Moroccan in Norway, mistakenly believed to be part of the Black September Munich Olympic massacre group. Six Israelis were arrested by Norwegian police and, after being jailed, were released short of serving their full sentences. The intended "target" was killed by a car bomb in Beirut in early 1979.

Ma'alot - 15 May 1974

Three Democratic Front for the Liberation of Palestine terrorists seized a school in northern Israel and four teachers and 100 children, and killed two Arab women and three Israelis. Two teachers and some children were released to convey demands for release of 26 Arabs in Israeli custody. Defense Minister Moshe Dayan ordered storming by a *Sayaret Matkal* force after negotiations failed. The terrorists and 22 children were killed, and sixty injured. Major political controversy erupted in Israel in the aftermath.

Scheveningen Prison - 31 October 1974

Four convicts seized 22 hostages, mainly visiting musicians. Dutch marines attacked using flares, sirens, and firing automatic weapons at the ceiling. Convicts were apprehended and hostages freed without casualties.

Moluccan Terrorist Train Siege - 2-14 December 1975

Hijacking of a train by South Moluccan nationalist terrorists and parallel seizure of the Amsterdam Consulate of Indonesia ended in surrender.

Entebbe Raid - 27 June-4 July 1976

In Operation *Jonathan* (named after raid commander Lt.Col. Jonathan Netanyahu), Israeli paratroopers released 106 Jewish hostages from Air France airliner hijacked by Arab and German terrorists. Twenty Ugandans, seven of eight terrorists, and Netanyahu were killed, and 11 MIGs destroyed on ground. Ugandans killed an elderly hospitalized woman in retaliation.

Moluccan Terrorist Train Siege - 23 May-11 June 1977

Royal Netherlands Marines killed six Moluccan terrorists and two of 80 hostages. SAS served as advisors. A parallel school hostaging ended in surrender.

Mogadishu - 13-18 October 1977

Four Palestinian terrorists hijacked a Lufthansa Boeing 737, and flew from Majorca to Rome, Cyprus, Bahrain, Dubai, and Aden, where the aircraft captain was killed, then on to Mogadishu, Somalia. They wanted to exchange a kidnapped west German industrialist and the passengers for $15 million and release of terrorists held in West Germany. In Operation MAGIC FIRE, a 26-man unit of the *Grenzschützgruppe 9*, with SAS assistance, using stun grenades that befuddled the terrorists, stormed the plane and killed three of them. Three German convicts committed suicide upon receiving news of the failed hijacking.

Chimoio - 23 November 1977

Raid by Rhodesian Special Air Service on a main encampment of the Zimbabwe National Liberation Army in Mozambique. The largest of several such incursions against camps in Tanzania and Mozambique until the end of white rule in Rhodesia in 1979. Approximately 1,200 Zimbabwean casualties; one RSAS trooper was killed.

Larnaca Airport - 25 September 1978

Two Palestinian terrorists, having killed a close political associate of Anwar Sadat, took 30 hostages at a major Cyprus hotel, released 19, gaining a Cyprus Airways DC 8, and cruised the Mediterranean while being denied landing rights in country after country. After refueling in Dubai, the plane returned to Lanarca Airport. A 75-man *Sa'aqa* (Egyptian commando) force landed close behind, but after exchanging heavy fire with the Cyprus National Guard for almost an hour and having 15 men killed and their C-130 transport destroyed, the survivors surrendered and returned to Egypt. A Cypriot court death sentence for the two terrorists was later commuted to life imprisonment.

Operation *Eagle Claw* - 24-25 April 1980

A raid aimed at freeing 52 American hostages held by Iranian supporters of the Ayatollah Khomeini, *Eagle Claw* was aborted when three of eight U.S. Navy helicopters malfunctioned; during departure, one crashed into a tanker plane on the staging site. Delta Force and other special operations elements were commanded by Colonel Donald Beckwith. Politically, the failure discredited the Carter administration, and both the Joint Chiefs of Staff system and special operations organization were heavily criticized. The hostages were released on the eve of Reagan's inaugural, and U.S. special operations refurbished.

London - 5 May-30 April 1980

Six Iranian Arab nationalists seized the Iranian embassy in Prince's Gate, London and held 26 Iranian and British hostages; five were released and two killed. A counterrevolutionary unit from B Squadron of the SAS entered, using special gear and stun grenades, killing five. The survivor received life imprisonment. Although some criticized SAS heavy-handedness, the action generated widespread acclaim, and the official march of the SAS became a hit record in Britain.

Barcelona - Banco Central - May 1981

The GEO - *Grupo Especiale Para Los Operaciones* - recently formed at that time from politically moderate volunteers from the Guardia Civil and other Spanish security elements, stormed the Banco Central seized by 24 gunmen holding over 200 hostages. Some terrorists got away in the fracas, but only one hostage was killed.

Padua - February 1982

The rescue without casualties of U.S. Brigadier General James Dozier from the Red Brigade by Italian police elite "Leatherhead" raiding unit was the first case in twelve years of a kidnap victim of the *brigatisti* being freed; former Primer Minister Aldo Moro had been killed after being held for almost two months in 1978.

Gaza - 12 April 1984

Four Palestinian Arabs hijacked an Israeli bus; Army elements killed two and the two remaining died after being turned over to the *Shin Bet*'s (counterintelligence service) antiterrorist unit. Photographs showed them unmarked when captured and a cover story that they had been badly wounded in the attack collapsed. While the press criticized the government, most of the Israeli public backed *Shin Bet*'s antiterror efforts.

Amritsar - 5-6 June 1984; 30 April 1986; 10-18 May 1988

The leader of Sikh militants, Jarnail Singh Bhindranwale, and about 500 of his followers and up to 700 Indian troops were killed as the latter attempted to occupy the Sikh Golden Temple and disperse the militants. Elite forces were committed initially, but tenacity of Sikh resistance led to major escalation, including direct tank-gun fire. Security forces, barefoot in deference to Sikh sensibilities, cleared the temple again on April 30, 1986, using shock grenades and light weapons, with one fatality. The political aftermath included a major surge in Sikh nationalism and terrorism, leading to the assassination of Prime Minister Indira Gandhi in October 1984.

Sikh activism continued, and the Golden Temple was besieged once more in May 1988, as Indian police special units killed and captured several dozen insurgents.

Auckland - 10 July 1985

A Greenpeace ship, *Rainbow Warrior*, about to lead an anti-nuclear demonstration fleet to the French Pacific nuclear test site was sunk at moorings by bombs placed by frogmen under French security services direction. Two agents of the *Direction Generale de la Securité Exterieure* received ten-year sentences from a New Zealand court. After an initial cover-up, the role of DSGE was made public. Its head was relieved, and the Minister of Defense resigned. After major diplomatic and economic friction between New Zealand and France, convicted DSGE agents were transferred to a French prison in the Pacific, and their status became a major element in New Zealand - French relations.

Achille Lauro hijacking - 7-11 October 1985

When an Italian cruise ship was hijacked by four Palestinians, Leon Klinghoffer, a U.S. citizen, was killed and thrown overboard, then terrorists surrendered to Egyptians. The Egyptian airliner carrying them to Tunisia was forced down at NATO base in Sicily. A U.S. Delta Force-Italian police confrontation ended when President Reagan ordered Delta to "stand down." The Italians captured, then released the terrorists. Craxi's coalition government fell in the aftermath.

Luqa Airport - 23-24 November 1985

Egyptian *Sa'aqa* commandos stormed a hijacked Egyptian airliner on Malta after hijackers released 13 women and began shooting hostages. One died. The *Sa'aqa* raiders blew a hole through the baggage compartment roof, but surprise was lost, and gunfire, grenades, and a resultant fire killed 57. Terrorist links were traced to both Egyptian Liberation Army and Palestinian factions.

Appendix D: Principal Airborne Combat Operations

Date	Location	Size of Airborne Unit Involved	Nationality of Airborne Forces
1933 (?)	U.S.S.R.-Central Asia	Company	U.S.S.R.
4/09/40	Denmark--several areas	Company and Battalion	German
4/09/40	Norway--Oslo; Stavanger	Company	German
4/18/40	Norway--Dombaas	Battalion	German
5/10/40	Holland--several areas	Division (+)	German
5/10/40	Belgium--Ft. Eben Emael	Company (+)	German
5/20/40	Norway--Bjornfeld	Battalion	German
2/10/41	Italy--Tragino Aqueduct	Platoon	British
4/25/41	Greece--Corinth Canal	Company	German
4/30/41	Greece--Cephalonia	Battalion	Italian
5/22/41-5/23/41	Crete	Division (+)	German
1/11/42	N. Celebes-Melado	Battalion	Japanese
1/42-2/42	Vyazma	Brigade	U.S.S.R.
2/14-15/42	Java--Palembang	Battalion	Japanese
2/20/42	Portuguese East Indies--Timor	Battalion	Japanese
2/27/42	France--Bruneval	Company	British
11/42	Algeria-Tunisia	2 Battalions	U.S.-British
7/3/43	New Guinea--Noemfoor	Regiment (-)	U.S.
7/10/43	Sicily	Division (+)	U.S.-British

Date	Location	Size of Airborne Unit Involved	Nationality of Airborne Forces
9/3/43	Salerno	Division	U.S.
9/24-25/43	Voronezh Front--Dnepr	Brigade	U.S.S.R.
3/5-12/44	Burma	Division (-)	British-Indian
6/6/44	Normandy	3 Divisions	U.S.-British
6/23/44	Philippines--Cagayan Valley	Regiment	U.S.
8/15/44	French Riviera	Division (-)	U.S-British
9/5/44	New Guinea--Nadzab	Regiment	U.S.
9/17/44	Eindhoven--Nijmegen--Arnhem	3 Divisions (+)	U.S.-British-Polish
12/6/44	Philippines--Buranuen (Leyte)	Battalion	German
12/19/44	Belgium--Ardennes	Battalion	Japanese
2/2/45	Philippines--Tagaytay Ridge	Regiment	U.S.
2/16/45	Corregidor	Regiment	U.S.
2/17/45	Philippines--Los Baños POW Camp	Company	U.S.
2/28/45	Germany--Breslau	Battalion	German
3/24/45	Rhine Crossing--Wesel	2 Divisions	U.S.-British
5/2/45	Rangoon	Battalion	Gurkhas
8/45	Manchuria	Detachments of various sizes	U.S.S.R.
3/23/51	Korea--Munsan	Regiment	U.S.
10/20/51	Korea--Sukchon--Sunchon	Regiment (+)	U.S.

Date	Location	Size of Airborne Unit Involved	Nationality of Airborne Forces
11/20/53	Laos--Dìènbìènphu*	Regiment (+)	French
10/56	Sinai--Mitla Pass	Battalion	Israel
10/56	Egypt--Port Said area	4 Battalions	British-French
6/67	Sinai--Mitla Pass	Battalion	Israel
10/83	Grenada	2 Battalions	U.S.

*Also: Numerous French and Vietnamese drops in Indochina and Vietnam from 1948-1954 (platoon to brigade size).

Author Index

Title Index

Subject Index

About the Author

Roger Beaumont is Professor of History at Texas A&M University. He is the author of *Military Elites: Special Fighting Units in the Modern World*, *Sword of the Raj: The British Army in India, 1747-1947*, and *The Nerves of War*, a short history of command and control. His numerous articles have appeared in publications such as *Journal of Contemporary History*, *U.S. Naval Institute Proceedings*, *Signal*, *Royal United Services Institute Journal* and *Military Affairs*. Dr. Beaumont is cofounder and North American editor of the journal *Defense Analysis*.